OUT OF THE DARKNESS
The Planet Pluto

OUT OF THE
DARKNESS
The Planet Pluto

Clyde W. Tombaugh
Patrick Moore

 STACKPOLE BOOKS
Harrisburg, Pa.

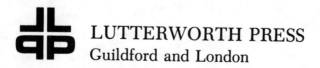 LUTTERWORTH PRESS
Guildford and London

OUT OF THE DARKNESS: THE PLANET PLUTO

Copyright © 1980 by
Clyde W. Tombaugh and Patrick Moore

Published by
STACKPOLE BOOKS
Cameron and Kelker Streets
P.O. Box 1831
Harrisburg, Pa. 17105
First printing, August 1980
Second printing, January 1981
Published simultaneously in Don Mills, Ontario, Canada
by Thomas Nelson & Sons, Ltd.

First published in Great Britain 1980 by
Lutterworth Press, Farnham Road, Guildford, Surrey.
ISBN 0 7188 2500 4

Printed in the U.S.A.

Library of Congress Cataloging in Publication Data

Tombaugh, Clyde William, 1906–
    Out of the darkness, the planet Pluto.

    Includes index.
    1. Pluto (Planet)   I.   Moore, Patrick, joint author.
II.   Title
QB701.T65   1980   523.4'82   80-36881
ISBN 0-8117-1163-3

Dedication

This book is dedicated to the former senior members of the Lowell Observatory, now deceased, who for so many years endured frustration, heartbreak, frugality, and paved the way to the final success of adding the ninth planet to the Solar System:

Roger Lowell Putnam, trustee

V. M. Slipher, director

C. O. Lampland, assistant director

E. C. Slipher, early searcher

Stanley Sykes, machinist

C. A. R. Lundin, optician (Alvin Clark firm)

under the inspiration of Percival Lowell, founder of the Lowell Observatory.

Also, to Patricia Tombaugh, who shared the early frugal situation and the loneliness of a husband who spent many long nights in the dome.

Contents

Foreword

The book you are about to read will take you on an epic journey of discovery. For two centuries astronomers have been searching and studying the outermost realm of the Solar System. Three planets—Uranus, Neptune, and Pluto—and their moons have resulted, and the stories surrounding these discoveries have a sense of devotion and mystery that leads all of us to look forward to the next striking discovery. Is it centuries away? Or just days away?

Discovery is where the scientist touches Nature in its least predictable aspect. It discloses to us the regularities of Nature, but in itself, discovery is fickle, striking at the unexpected moment. This is the view that I must take after my serendipitous discovery of the moon of Pluto. Certainly the manner in which the stage is set for a discovery is normally accomplished with scientific imagination and discipline, as in Clyde Tombaugh's discovery of the planet Pluto. However, the moment of discovery is always of another sort, a moment of stepping across an unknown threshold into uncharted waters. In the moment of discovery something in our soul reaches out to Nature and reaps a return in kind. Let me tell you of my personal experience with discovery.

In the summer of 1977 I had initiated a request for the astronomers of the Naval Observatory's Flagstaff Station to obtain several photographs of the planet Pluto. These photos (which are called plates because the photographic emulsion is on flat, glass plates to ensure dimensional stability) were to be shipped to Washington where we would measure the position of Pluto with high precision on our measuring machine called Starscan. This position would be used to improve the accuracy of the orbital elements of Pluto's motion. In fact, some fifty plates of Pluto from the years 1965, 1970, and 1971 had already been measured for this same purpose. As a result of this request, plates were exposed with the 5.085-foot (1.55-meter) Astrometric Reflector in Flagstaff, Arizona on 13 April, 20 April, and 12 May 1978. These observations from 1965 to 1978 set the scientific stage for a most unexpected discovery.

On Thursday morning, 22 June 1978, I began to measure the recent Pluto plates on our Starscan measuring machine. During the following week I would take off work in order to move my family from our apartment into a home. Because most of the moving would be done by myself in Washington's summer heat, I found myself unusually relaxed in the cool and dark measuring room, content with anticipation of the effort to come.

The plates I was measuring had been considered poor because the images appeared defective. There were two plates from each date and three exposures on each plate; the plate had been moved slightly between each 1.5-minute exposure, a procedure which saves on plates and allows easy comparison of the three side-by-side images. As I moved the machine from image to image, they were projected thirty times enlarged on the 3-foot (.91-meter) screen in front of me. It was true that the image of Pluto was about 30 percent elongated in a north-south direction. This was sufficiently bad to consider the plate rejected for measurement, because such asymmetric images result in false positions. However, I proceeded to measure the reference star positions; the position of Pluto would be derived from the already known positions of other stars on the plate. These were not badly elongated, so I continued and completed the measuring of one plate on each of the three dates.

After the first and third of these plates I typed on the computer record: "PLUTO IMAGE TRAILED IN DEC, REFERENCE IMAGES NOT TRAILED." DEC refers to declination, the north-south direction in astronomical coordinates. At this point some computer maintenance men arrived on the scene, and I had to interrupt my measuring. I was confused by the failure of the Pluto image to show the same shape as the star images on the same exposure, so I put the plates, one by one, under a microscope where I could better compare the six plates. It was clear that each of the six Pluto images on 13 April and 12 May suffered from an elongation not shared by the reference star images; the 20 April images were of lesser quality and the elongation somewhat ambiguous.

The elongation was fainter than the core of Pluto's image, which made it appear unlike an image caused by motion of Pluto in its orbit. I thought about the possibility of a flare of some sort being emitted by Pluto, but at the instant I realized that the 12 May elongation was north of Pluto and that the 13 April elongation was to the south, the concept—moon—jumped into my thoughts.

Being familiar with disappointments, I was more incredulous than excited by the thought. I asked my colleague, F. Jerry Josties, to look at the plates. Jerry arrived at a similar conclusion. It was nearly noon; we spent the rest of the afternoon checking out alternative explanations. Neither motion of Pluto, background stars in chance juxtaposition with Pluto, or defective images offered a plausible alternative. We concluded the day believing (and mentioning to other staff members) that Pluto had a moon.

Although Thursday was a day of discovery, Friday—23 June—was to provide a second, equally surprising discovery. I examined the fifty plates in our collection of Pluto observations. On most dates, the seeing conditions (image blur as light passes through our atmosphere) were not sufficient to allow detection of even a trace of elongation. However, by remarkable good fortune, five plates during a single week of June 1970 had detectable elongation. The elongation angle proceeded clockwise around Pluto during the week of exposures; it was easiest seen in the north-south direction but less so at other angles.

Thus, within twenty-four hours of the discovery, I was able to hypothesize that the moon had a six-day orbital period with the orbit inclined to the line-of-sight, such that the moon was most likely to be seen every third day at the north and south elongations.

Furthermore, because Pluto's brightness was known to vary with a period of 6.3867 days (details are given in the text), we assumed the period of the moon to be exactly the same. My colleague, Dr. Robert S. Harrington, suggested that I make careful estimates of the elongation angles. Meanwhile, he computed an ephemeris using the 6.3867-day period and worked backward in time from the angle on the plate of 12 May 1978. Harrington's predicted angles coincided almost exactly with my estimates, thereby verifying that an orbit existed. Thus, in two days, we had discovered a moon and an orbit—and I was free to start moving furniture.

During the move, I frequently thought about a name for the new moon. Although my early favorites were Oz, from *The Wizard of Oz*, and Charon, after my wife, Charlene—Char to her family; I soon learned that the astronomical community expected the name to be taken from Greek mythology and to be associated with the god Pluto. But serendipity struck again, and I made my third discovery of the week as I opened my dictionary and read, "Charon—in Greek mythology, the boatman who ferried dead souls across the river Styx to Hades, the domain of Pluto." So be it; ours is not to reason why—at least not immediately. Charlene and I were delighted with Charon; despite his somewhat grizzly reputation, he appeared to be doing essential work. Several months later, Bob and Betty Harrington named their daughter Ann Charon Harrington, so part of this story is just beginning.

The discovery of Charon required excellent equipment, the Astrometric Reflector and the Starscan machine; it required dedicated observers, in particular, Dr. Anthony V. Hewitt, who obtained the discovery plates of 13 April and 12 May 1978; and it required several nights of unusually stable atmosphere which coincided with times of maximum elongation of Charon, a gift of Nature. Discovery is fickle. The Charon plates were exposed just 4 miles (6.4 kilometers) from Lowell Observatory, the site of the discovery of Pluto by Clyde Tombaugh; and I found the image of Charon on Starscan, just 100 yards (91.4 meters) from the 26-inch (66.04-centimeter) refractor, the site of the discovery of the moons of Mars in 1877 by Asaph Hall.

This book will tell you more of the precision, labor, passion, and wonder that is astronomical discovery. In Greek myth, souls were sometimes kept waiting many years on the shores of the river Styx until Charon took them across into the afterlife, the domain of Pluto. Today, Pluto and Charon still stand guard at the boundary of the Solar System, challenging us to discover, daring us to draw new mysteries out of the darkness.

JAMES W. CHRISTY

Preface

This book tells the story of the discovery of the ninth planet, Pluto, from my own personal experience. I have attempted to relate a complex scientific event in as popular language as much as possible. The technical aspects of an astronomical problem are expressed in descriptive form and illustrated with diagrams and photographs to aid the reader in grasping the fundamental concepts involved. Deliberately, I have not included a single mathematical equation.

The choice of dimensional units to be used has presented a dilemma. The United States is in an early stage of transition to the metric. At the present time, the majority of readers in the U.S. do not visualize metric dimensions. Hence, the units are expressed in the English system. Readers in foreign countries have long used the metric system. Since continual reference to footnotes or to an appendix would make for awkward reading, it seemed best to give the first number in the English system followed immediately by the metric equivalent within parentheses.

The success in finding the new planet was not due to complex mathematical theory, but to basically simple observational procedure and an enormous amount of painstaking work. Contrary to widespread opinion, the mathematical prediction was of little aid in actually finding the planet, because of earlier negative observational results. Singling out the planet from a sky background, teeming with millions of stars, was like finding a needle somewhere in a large haystack.

No other story about Pluto could be written like this one. It is flavored with human interest and amusing incidents. Several aspects of the problems involved in the planet search project have not been previously described to the public. Indeed, the scientific community is not aware of some aspects, judging from expositions in textbooks and others. One of the questions so frequently asked, "How did you know the object was beyond the orbit of Neptune when you first saw it?" Indeed, several professional astronomers did not understand the key strategy of the planet search. The reader will perceive how simple in geometry it really was. But the searching through the millions of star images was hard, tedious, careful work, which taxed my perseverance to the limit. Inspecting the photographs of some of the star fields will illustrate the nature of the task.

My co-author, Dr. Patrick Moore of England, a writer of several popular books on Astronomy, has written five of the chapters. This was appropriate because England was much involved in the discoveries of the planets Uranus and Neptune. Pluto is the only planet discovered to date in America. The few planet discoveries were accomplished over a span of 150 years and were rare events.

These planet discoveries have involved blunders, frustrations, remorse, controversies, and triumphs in several peoples lives. Thus, these historical events provide more than just scientific tedium. They furnish all of the drama essential to a good novel, one that actually happened. If some of the passages seem overly technical to some readers, skip over those pages and go on.

Reflecting upon this whole episode in later years, I came to realize that the discovery of Pluto was due to a remarkable chain of accidental events spanning several decades, decreed by fate. Percival Lowell and his younger assistants struggled against great odds, but they did not give up, in spite of the widespread skepticism in the astronomical world.

A new planet, if there was one, could be anywhere in the Zodiac belt, or even outside it.

After much careful planning on a frugal budget, the Lowell Observatory completed the wide-angle 13-inch (33-centimeter) telescope, superbly suited to resume the planet search in early 1929. This is where I came upon the scene, under very unusual circumstances, a farm boy amateur astronomer without a university education, to find the faint planet, Pluto, thirteen months later. Almost immediately, Pluto presented problems as to its status and touched off lively debates and controversies. The discovery captivated the world and was voted by the Associated Press as one of the top ten news events in the world for the year 1930.

At the time of discovery, Pluto was nine times the Earth's distance from the Sun beyond the orbit of Neptune. Except for the uniqueness of its orbit, little was known about Pluto. It did not seem to fit into any known class of objects in the solar system. Accordingly, there was much controversial speculation regarding its nature and origin. For the past fifty years, Pluto has been moving inward toward its perihelion point, just inside of Neptune's orbit. Its distance has decreased about 800 million miles, resulting in a relatively small, but appreciable, observational advantage.

In 1965, an upper limit in Pluto's size was established by the near miss of Pluto occulting a star. In 1976, special infrared observations indicated that Pluto's surface was covered with methane frost. This raised the albedo value and drastically reduced the diameter or size. In 1978, a very close, but relatively large satellite was discovered for Pluto, providing the first reliable value for the mass of Pluto—shockingly small. This immediately yielded the surprisingly low density of about 1.0, indicating that Pluto and its satellite, Charon, are two huge icebergs. Many of the old speculations were put to rest, but new speculations were generated.

Thus, Pluto has perpetuated a strong continuing interest. Its nature is more strange than ever. Its status as an object is engulfed in mystery. Everything about Pluto was unexpected. One can only speculate on what new things will be learned about Pluto in the future.

February 18, 1980, marked the fiftieth anniversary of the discovery of Pluto.

It is a pleasure to acknowledge the help of the following persons in the preparation of this book: Barbara McClanahan Tombaugh, for generous assistance in the preparation of the manuscript; Patricia Tombaugh, for helpful suggestions and encouragement; Chuck Williams and John Womack of New Mexico State University, for the preparation of photographs for illustration; and Dr. Herbert Beebe, head of the Astronomy Department at New Mexico State University, for encouragement and counsel. To my co-author, Patrick Moore of England, for much cooperative correspondence regarding the subject material, and the chapters that he wrote.

CLYDE W. TOMBAUGH
Professor of Astronomy Emeritus
New Mexico State University

In 1930, when I was a boy of seven, I attended a preparatory school in my native Sussex. Science classes were held once a week, and during one of these I first learned of the discovery of a new planet beyond Neptune. At that time, and for many years afterwards, I could have no idea that I would be invited to collaborate in a book with the great astronomer who made the discovery.

Pluto is not an ordinary planet; it has peculiarities all its own, which we do not yet fully understand. Faint though it may be, Pluto is of exceptional interest, and a book devoted to it has been long overdue. Only one man could do justice to the subject: Clyde Tombaugh himself.

Here, then, is the story of Pluto. My part in the book has necessarily been very minor. I can only add that it has been a tremendous honor and privilege to work together with a man whose achievements in science will never be forgotten. I am indeed proud to have had the opportunity of doing so.

PATRICK MOORE

1

An Autobiographical Sketch

by Clyde W. Tombaugh

I was born on a farm near Streator, Illinois, on 4 February 1906. My parents were Muron and Adella Chritton Tombaugh; both are now dead. I was the eldest of six children. After me came Esther, Roy, and Charles. On the farm we raised corn, oats, and hay. In 1922, we moved to a wheat farm near Burdett, Kansas (northeast of Dodge City). This is where the last two children were born: Robert and Anita Rachel.

In grade school, I loved geography and history. One day, while in sixth grade, the thought occurred to me, *What would the geography on the other planets be like?*

My Uncle Lee, who was somewhat of an amateur astronomer, lead me to some answers. He and his family lived on a farm near us, and we visited together frequently on the weekends. He had a 3-inch (75-millimeter) telescope with a magnifying power of 36 diameters. This means that an object such as the Moon appears 36 times wider through the telescope than it does to the unaided eye. The telescope was 3-feet (0.9 meters) long. It had a simple objective lens, which gave much spurious color around each object. All one could see were the larger craters on the moon, the small disk of Jupiter and its four largest moons, the ring of Saturn, and the phases of Venus.

My Uncle Lee loaned this telescope to me for several weeks at a time, as well as a very small, popular book on astronomy. I read and reread this book so many times that I could recite many pages, word for word, from memory, including all of the statistical figures of the planets, the Sun, and several stars, which I never forgot. In this little book, I learned about Galileo, Sir William Herschel, and Percival Lowell. These men became my boyhood heroes.

During the last year of World War I and for a few years follow- ing, there was a shortage of farm manpower. I had to help my father plant corn in May, cultivate the corn in June, and cut oats and wheat in July. From the age of twelve on, I spent the whole month of August taking a man's place on the threshing crews. I handled hundreds of tons of grain with scoop shovel and pitchfork. My father was the engi- neer on the threshing crew, and I would often sit on the steam engine with him while waiting my turn to pitch a load of bundles into the threshing machine.

One day, one of the newcomers to the threshing gang asked my father, "Where is your young brother?" "What brother?" my father asked. "That young feller who often sits on your steam engine," the man said. "Oh, that's my eldest son," my father said. The man shook his head in astonishment. He just could not fathom a father and son being such pals. Several times, my father would tell this story with much glee.

It required a crew of about twenty-five men to keep a large thresh- ing machine busy. A large number of farmers would go in together to compose a threshing run. It was the most exciting time of the year— delicious meals of food for the men at noon, wrestling matches, fights, snakes in the bundle shocks, and spectacular runaway horses. The noise of the machinery was music to my ears. Well, that all went out of style many years later with the advent of combines.

I learned quite a bit about steam engine operation and mainte- nance from my Dad. Cars did not interest me much, but steam engines and telescopes fascinated me.

Many of the farmers did not own their farms. Renting terms were often difficult. My father had to pay cash at a rate of so many dollars per acre, whether he had a good crop or not because of weather. So he decided to try his luck on a farm in southwest central Kansas in the big wheat country.

My family moved to Kansas in August 1922. A great deal of urgent work had to be done in preparing over 250 acres of ground for the fall seeding of wheat. It was necessary for me to drop out of high school for one year to assist my father in this work.

Dad and Uncle Lee had purchased a 2¼-inch (57-millimeter), 45-power achromatic telescope from Sears-Roebuck Company in 1920. When we moved to Kansas, my uncle gave up his share in it. Western Kansas had a drier and less cloudy climate than Illinois. During seasons of less pressing farm work, I spent hundreds of hours in the early night exploring the skies with this telescope. I learned to identify the stars and constellations thoroughly, and observed the motions of the planets. I read everything on astronomy that I could get my hands on. I studied my Dad's books on trigonometry and physics, dabbled in some Latin, and even learned to read some Greek. Studying at night was done beside a kerosene lamp as, in those days, farms did not have electricity.

I was not a bookworm altogether. I set up a football field in the pasture, and a bunch of the neighborhood boys would get together to play touch football on Sunday afternoons. Other times we played tennis on a court I fixed up in the backyard. Sometimes, we would high-jump, long-jump, pole-vault, and put the shot with a 12-pound (4.5 kilogram) steel ball from an old windmill. Later, I participated in these events on the track and field team during my last two years in high school.

About the first of May, 1925, our high school principal, Mr. Rigby, took the three top students of our senior class to a state scholastic meet where a few hundred of the top high school students of the state met in stiff competition exams. It was held on the campus of Kansas State University at Manhattan. (Later, I learned that I won fourth place in physics.) This was my first real contact with a university. I liked what I saw. I realized then that the world of academia was where I belonged. If only someday I could become a college professor. But how? Where? When? In my situation, such a goal seemed utterly impossible.

I graduated from the small high school in Burdett, Kansas, in 1925. There was no prospect of going on to college. Times were hard because of low crop yields. My father could not afford a hired man for the farm work.

By then I had become obsessed with getting a more powerful telescope. Buying one was totally out of reach. In 1924, I subscribed to *Popular Astronomy*, in which Latimer Wilson of Nashville, Tennessee, had published a paper entitled "The Drift of Jupiter's Markings," illustrated with four beautiful drawings he had made using his homemade 11-inch (28-centimeter) reflecting telescope. This really fired me up, so I wrote to him, asking how to make such a telescope and where one could obtain the necessary materials. He referred me to two issues of a boys' Sunday school paper in which he wrote an article on directions and gave addresses where the glass disks, grinding abrasives, pitch, and rouge for polishing could be obtained. They did not have complete mirror-making kits in those days. It was necessary to order each one from different places. He had recommended an 8-inch (20.3-centimeter) mirror for a starter. The disks were cut from thick, ship porthole glass.

In February 1926, I set up a grinding stand on a sturdy post on the south side of the house and proceeded with the grinding. After using the finest grade of abrasive, I melted pitch to make the polishing lap and polished with rouge. The Foucault knife-edge test—to determine the accuracy of the curve of the mirror—was set up in the house, but the heated house made the test very difficult and uncertain. After an unsuccessful attempt at silvering, I sent the mirror off to a French telescope maker, Napoleon Carreau, in Wichita, whose ad was in the *Popular Astronomy* magazine. He returned the mirror, silvered, saying that the figure was not very good. In the meantime, I constructed a tube out of pine boards and a mounting from discarded farm machinery. Setting circles were made on wood disks, laid out from a protractor I had used in geometry class. The setting circles are centered on the two telescope axes to measure angular distances in the sky. With the setting circles, I easily picked up the great globular star cluster in Hercules. The light power was sufficient to partially resolve it into stars. To view such a compact swarm of stars, at a distance of 20,000 light-years overwhelms one with a sense of the Eternal.

I had hoped to see some of those wonderful features on Mars during the fall opposition in 1926. But the mirror's figure did not permit useful magnifying powers above 84. This was a keen disappointment.

However, I was able to see the white south polar cap, and the larger, dark areas, called *maria*.

The directions had stated that it was better to test mirrors in a cellar, where the temperature would be much more constant, and free from air currents. Well, we did not have a cellar.

It was customary in that part of the country to build an underground cellar with heavy concrete walls and an arched ceiling near the house for two purposes: to store food, such as milk, butter, eggs, and home canned food to keep them cool in summer and prevent them from freezing in winter; and to provide a safe haven when very strong windstorms or tornadoes were imminent. We badly needed such an underground cave, so we planned to build one.

I persuaded Dad to build a long one, so that it would provide a long testing alley for the Foucault knife-edge testing of mirrors at center of curvature (twice the focal length of the mirror). He agreed, if I would dig the pit. After the wheat was drilled in September and the sorghums were harvested in early October, we were ready to work on the poured concrete cave project.

Well, we needed money to buy the large amount of sand and cement, so Dad hired out as a carpenter to a neighbor who was building a new house. With pick and shovel, I dug a rectangular pit 24 feet long by 8 feet wide by 7 feet deep (7.3 meters long by 2.4 meters wide by 2.1 meters deep). I dressed the sides smooth and straight so less lumber would be required for the concrete forms. Then I loaded wagonload after wagonload of the dirt and hauled it away. My Dad and I spent a few days building the forms which had to be heavily braced to withstand the pressure of the concrete.

Everything was made ready for the big day of pouring concrete. Some neighbors came over to help. With hoes and shovels, we mixed up about 12 cubic yards (9.15 cubic meters) of concrete and carried it by shovel to dump in for the walls. About three days later, the neighbors came again, and we mixed up at least 8 cubic yards (6.1 cubic meters) more of concrete to make the floor and arched roof, all steel reinforced. After a couple of days, we removed the lumber forms. Dad went back to work on the neighbor's house, and I proceeded to dress up the interior with a finishing coat of concrete. Then we built a concrete stairway and put in small windows at each end of the arched face.

This cave provided me with much improved conditions in which

to build better telescopes. My next project was a 7-inch (17.8-centi-
meter) reflector telescope for my Uncle Lee in Illinois. This time my
directions came from the first edition of *Amateur Telescope Making*,
which was published by the *Scientific American* and cost $2.00. I
ordered a copy and began studying it. The directions were much more
detailed than the ones I used before, and I learned where I had made
some mistakes on my 8-inch (20.3-centimeter). I ordered the materials
and made grinding and polishing stands. I began work in February
1927. The Foucault testing conditions in the cave proved excellent.
The telescope with mounting was finished in May and worked quite
well. Then a special treat came: Pons-Winnecke Comet became visible
in the evening sky. That was an interesting sight in the 7-inch
(17.8-centimeter) telescope. It was my first view of a comet. I shipped
the telescope to my uncle the following month.

With this money, my uncle sent me for making his telescope, I
promptly ordered glass disks and grinding supplies to make a 9-inch
(22.8-centimeter) telescope with a 79-inch (2-meter) focal length. This
time I was determined not to be denied some good views of Mars in the
opposition (exactly opposite the Sun) late in 1928.

During a short rainy spell in September 1927, I ground the curve
on the 9-inch (22.8-centimeter) mirror. The whole job was done in
twenty-four hours of work, including clean-up after each grade. The
cave was a marvelous workshop.

Late in the fall of 1927, I began polishing the mirror, and during
the late winter months of 1928, I fought the figure for a high accuracy.
Finally after many weeks, I got a superb parabolic figure. I sent it to
Carreau in Wichita for silvering. He was impressed with the figure.
The 9-inch (22.8-centimeter) gave sharp images with high magnifying
powers, up to 400 diameters. The views of the Moon and planets were
superb.

On June 20, 1928, we were the victims of an agricultural calam-
ity. An ominous, bluish-black cloud rose up from the northwest in the
late afternoon. We got the horses and cows into the barn. A terrible
hailstorm struck. It lasted about fifteen or twenty minutes. The noise
was deafening. Heavy rain followed. After it passed, Dad and I went
to the fields to inspect the damage. We had a twenty acre field of oats
just west of the house. It was shoulder high and would have been the
best we had ever raised. In a few more days it would have been ripe

Clyde W. Tombaugh, at the age of twenty-two, with his homemade 9-inch (22.9-centimeter) reflecting telescope on the family farm near Burdett, Kansas. This picture was taken in the summer of 1928, eighteen months before the discovery of Pluto.

enough to cut. The straw stalks were all completely beaten down into
the mud. Elsewhere on the farm there was not a spear of wheat stand-
ing. We would be flat broke until next year. I said, "Farming is not for
me. The first chance I have of getting out of it, I'm going." The prospect of going to college seemed more remote than ever.

The following weeks, I hired out to some neighbors a few miles
away where the hailstorm did not hit. I ran one of their combines, cutting several hundred acres of wheat. It was this harvest money I earned
that paid for my train trip to Flagstaff five months later.

I started making careful drawings of the markings on Jupiter and
Mars in October, and continued through November and December
1928. It was thrilling to watch these planets slowly rotating on their
axes as the markings drifted across their disks. Now I really wanted to
be an astronomer in an observatory. But without a university degree, I
had no credentials.

I was in a seething dilemma. My younger brothers were getting
old enough to help with the farm work. I was twenty-two, and it was
time to think of leaving the home roost. Two choices entered my mind:
set up a little telescope-making business of my own or apply for an ap-
prentice fireman's job on the railroad, hopefully some day to become a
railroad engineer.

In those days, no young man in his right mind would think of
marrying until he had an adequate and secure job. I didn't want just
any old job, for I observed around me that those who did were likely
to be stuck with that kind of work for life. I had to choose carefully.

In the November and December issues of my old 1924 copies of
Popular Astronomy, I had read observation reports on Mars written by
the Lowell Observatory staff. It was the only planetary observatory I
knew of. I sent several of my drawings of Jupiter and Mars to them.

About that time, I received a letter from Napoleon Carreau, saying he was getting more orders for telescopes and asking if I would be
interested in coming to Wichita "sometime" the following year to help
him.

In the meantime, events were taking place at Flagstaff, of which I
knew nothing, that would change my life completely.

Sending my drawings of Mars and Jupiter to Dr. V. M. Slipher,
director of the Lowell Observatory, could not have been better timed.
Slipher was looking for a suitable amateur to work with a newly acquired 13-inch (33-centimeter) telescope. He responded almost imme-

diately with a letter, asking a number of questions, which I promptly answered. Then came another letter with more questions, one of them being, "Are you in good physical health?" which again was promptly answered. I realized that there was more than just polite interest.

Then came the final letter, in which he stated that they were about to put in operation a new photographic telescope. He said that the night work would involve long exposures in an unheated dome. "Would you be interested in coming to Flagstaff on a few months trial basis, about the middle of January?"

I was thrilled at the prospect of such an extraordinary opportunity. Since the telescope-making job in Wichita was not firm, there was no problem of a decision. So I sent my final letter stating when I would arrive in Flagstaff.

The next week, I made preparations for the long journey. My mother selected the clothes to take, and I chose several of my astronomy books and high school physics and math books.

On the appointed day of departure, 14 January 1929, it was hard to say good-bye. My mother had tears in her eyes. How long would it be before we would see each other again? My father took me to Larned, thirty miles (48 kilometers) away, to board the train. I bought a one-way ticket, but could not afford a Pullman for the twenty-eight hour trip. My father's parting words were, "Clyde, make yourself useful, and beware of easy women."

I stepped aboard carrying my heavy suitcase and found a seat beside a window. As the train moved out and we waved through the window, a lump came in my throat. For a few minutes I felt very sad, almost wishing that I wasn't going.

As the train picked up speed, I thought to myself, *Cheer up, there will be a lot of new and beautiful scenery to see.* I had never been to the far Southwest. I had some cause for apprehension. Tomorrow, I would be a thousand miles from home. I did not know a single soul in Arizona, except for Dr. Slipher by his letters, and I did not have enough money in my wallet for a return ticket.

I decided to write a travelogue of the trip for my first letter home.

By late afternoon, the Spanish Peaks loomed into view, my first view of the mountains since 1919. By evening, the train reached Trinidad and stopped for dinner. I had sandwiches that my mother had prepared.

Before starting out again, extra engines were attached for the steep climb up over Raton Pass. I had never heard such engine puffing. It was dark, and coal cinders from the smokestacks fell like hail on the passenger cars. I gazed out the window to take in everything that I could see.

After Raton Pass, the train moved very fast again to Las Vegas. Then the squealing of the wheels around the sharp curves of the narrows near Lamy. Then Albuquerque. I dozed off, awakened by daylight as we approached Gallup. Enormous red cliffs for miles and miles. They looked ten times higher than they appear to me now. We stopped at Gallup for breakfast. At Winslow, another locomotive was put on the train, for soon it would begin the 2,000-foot (610 meter) climb to Flagstaff. In a little while the train slowed down to cross the trestle over Canyon Diablo, deeply cut in a treeless semidesert. I could see the Painted Desert to the north of me. To the west loomed the San Francisco Peaks, the highest mountains in Arizona. Then appeared the small junipers, then the pinions, and finally the tall ponderosa pines on the snow-covered plateau. For over an hour, the engines labored hard. Around the curve, Flagstaff came into view. It was after one o'clock. As I stepped off the train, the conductor handed my suitcase to me. "Quite a little trunk you have there, sure is heavy, what do you have in it?" he asked. "Books," I said.

I entered the depot, looking lost and bewildered. A gray-haired man stepped forward. "Are you Mr. Tombaugh?" he asked. "Yes," I replied. "Glad to see that you arrived safely. I am Dr. V. M. Slipher, who wrote the letters to you."

I have often wondered what his first impression was of this raw, young man from the plains of Kansas, whom he had recruited for his staff, and whose discovery only thirteen months later would rock the astronomical and academic world and the world in general.

2

Setting the Scene

by Patrick Moore

Far away in space, so remote that it takes more than 164 years to complete its journey round the Sun, there orbits the giant planet Neptune. It has been known ever since 1846, and was once believed to mark the outer boundary of the main Solar System. Nowadays we know that this is not true. We have found another world, Pluto, which is even more lonely and desolate.

Pluto has been nicknamed "the puzzling planet." It seems to be in a class of its own; its orbit is unusual; and its makeup is probably different from any other member of the Sun's main family. It may even be that Pluto does not deserve to be ranked as a true planet at all. Detailed studies of its surface are impossible as yet; a telescope of fair size is needed to show it at all, and its apparent diameter is too small to be measured accurately. A few nights ago I uncapped the 15.3-inch (39-centimeter) reflector in my home observatory, and turned it toward Pluto. There was the planet—looking exactly like a dim star.

Pluto was discovered in 1930 by Clyde W. Tombaugh, the senior author of this book. The discovery was not due to chance. Tombaugh was carrying out a deliberate search, using calculations made more than a decade earlier by Percival Lowell, founder and first director of the famous observatory at Flagstaff, Arizona. Whether or not Lowell's calculations were valid is still a matter for debate, but in any case they led to the right result. Pluto was very close to the position which Lowell had predicted. Coincidence? If so, it was certainly an amazing one.

Since Lowell plays an all-important part in the whole story of Pluto, we must begin by saying something about him. He had a varied career, but his main love was astronomy, and the setting-up of the observatory at Flagstaff was entirely his doing. He was particularly interested in the Red Planet, Mars, and he carried out a long series of observations which ended only with his death in 1916. For most of his work he used the fine 24-inch (61-centimeter) refractor at Flagstaff—one of the larger telescopes of its kind and certainly one of the best in the world. Unfortunately, Lowell was wedded to the idea that the famous Martian "canals" were artificial waterways built by intelligent beings in an effort to conserve every scrap of water upon their dried-up planet. Theories of this kind met with strong opposition even in Lowell's lifetime, and we now know that they are completely wrong. There are no canals on Mars, and there are no Martians. Moreover, the dark areas on the planet once thought to be old seabeds filled with vegetation are not due to organic material. As yet we have no proof of any living thing on Mars.

In the popular mind Lowell is always associated with his belief in a Martian civilization—not for his outstanding work in other directions. This may be natural, but it is grossly unfair. Perhaps the present book will do something to put the record straight.

First, then, let us do our best to set the scene and to put matters in some sort of perspective.

The Sun is a star. It is by no means exceptional, and modern astronomers even relegate it to the status of a yellow dwarf. Many of the stars visible on any clear night are much larger, hotter, and more luminous than the Sun, but they are also much further away. Even the closest of them, Proxima Centauri, is so remote that its light, moving at approximately 186,300 miles (300,000 kilometers) per second, takes over four years to reach us. The Sun, on the other hand, is a mere

92,957,000 miles (150,000,000 kilometers) away on average, so that its light can reach the Earth in only 8.6 minutes.

Of the nine known planets, Mercury and Venus are closer to the Sun than the Earth is; Mars, Jupiter, Saturn, Uranus, Pluto, and Neptune are further away. Note that I have listed Pluto ahead of Neptune, because at the present time it is moving within Neptune's orbit. Its mean distance from the Sun, however, is much greater. The principal data about the planets may be conveniently summarized in a table.

Planet	Distance from Sun (in millions of km.)			Revolution Period	Rotation Period	Equatorial diameter (km.)	Number of Satellites
	Min.	Mean	Max.				
Mercury	45.9	57.9	69.7	88 d	58.6 d	4,880	0
Venus	107.4	108.2	109.0	224.7 d	243 d	12,100	0
Earth	147.1	149.6	152.1	365.26 d	23h 56m	12,756	1
Mars	206.7	227.9	249.1	687 d	24h 37m	6,787	2
Jupiter	740.9	778.3	815.7	11.86 y	9h 50m	142,800	15
Saturn	1347.	1427.	1507.	29.46 y	10h 14m	120,000	13
Uranus	2735.	2870.	3004.	84.0 y	± 23h	51,800	5
Neptune	4456.	4497.	4537.	164.8 y	± 22h	49,500	2
Pluto	4425.	5900.	7375.	247.7 y	6d 9h 17m	3,000?	1

d = days, y = years, h = hours, m = minutes

Any map of the Solar System will show that it is divided into two main parts. The inner group of planets (Mercury to Mars) is made up of solid, relatively small worlds, while the outer giants (Jupiter to Neptune) have gaseous surfaces. Between these two groups there is a wide gap, in which move thousands of dwarf worlds known variously as the asteroids, planetoids, or minor planets. Of these, only Ceres is over 620 miles (1000 kilometers) in diameter, and only Vesta is ever visible to the naked eye.

What, then, of Pluto? It is certainly small, and now seems to be even smaller than Mercury or some of the satellites of the giant planets. Pluto's mean distance from the Sun is much greater than that of Neptune, which is why Pluto is known as the trans-Neptunian planet; it was well beyond Neptune when it was discovered in 1930. However, greater orbital eccentricity means that part of its orbit lies within that of Neptune, and between 1979 and 1999 the distance of

Pluto from the Sun is less than that of Neptune. There is no danger of collision, because Pluto's orbit is inclined at the unusually sharp angle of 17 degrees. Indeed, at the present epoch, Pluto can approach Uranus more closely than to Neptune. It next reaches its closest approach to the Sun, its perihelion, in 1989, so that between 1979 and 1999 it temporarily forfeits its title of "the outermost planet."

The five naked-eye planets must have been known since the dawn of human history, even if their nature was not appreciated. By Greek times it had become clear that they were quite different from the so-called fixed stars; they wandered slowly around the sky, even though they kept strictly to one particular band, the Zodiac. The Zodiacal band is the strip of sky to either side of the Ecliptic (the yearly path of the Sun among the stars). Of the planets, only Pluto can move beyond the Zodiacal band. Together with the Sun and Moon, there were seven members of the Solar System. This seemed logical enough, since seven was the mystical number of the ancients. Aristarchus of Samos, one of the most enlightened of the Greek astronomers, went so far as to suggest that the Earth itself was in orbit around the Sun, though this bold idea was not finally accepted until many centuries after Aristarchus's death around 230 BC.

By the end of classical times, the apparent movements of the naked-eye planets were well known, and tables of remarkable accuracy had been drawn up. The Arabs of a thousand years later improved these tables considerably, but before the invention of the telescope there was no way of learning much about the planets themselves. In January 1610, Galileo, the greatest (though not the first) of the early telescopic observers, discovered the four bright satellites of Jupiter and established that Venus shows lunar-type phases. He also found that Saturn had a decidedly unusual shape, though his low-powered "optick tube" was not strong enough to show the ring system in its true guise. Subsequently, with the improvement of telescopic equipment, planetary observation made great strides. Then, in 1781, William Herschel discovered Uranus moving well beyond the path of Saturn. It would be churlish to belittle this discovery, but it is true to say that Herschel was not looking for a new planet and did not at once recognize it even when he had found it. He believed it to be a comet, and not for some time was its planetary nature established.

Uranus is dimly visible with the naked eye to a keen-sighted observer who knows where to look for it, but the two remaining members of the solar family, Neptune and Pluto, are much fainter. Both were tracked down by mathematical calculations, so that in a sense they were "discovered" before they were actually seen.

Mercury, aptly named after the swift-moving messenger of the gods, is never very prominent because it keeps to the same region of sky as the Sun, and cannot be seen against a dark background. It is relatively dense, with an iron-rich core and a weak, but detectable, magnetic field. From Earth, little can be seen on its surface even with powerful telescopes, but in 1974 the United States' probe, *Mariner 10*, made the first of three active passes and sent back excellent pictures from close range. Mercury has mountains, ridges, valleys, basins, and craters; superficially its surface looks very like that of the Moon. There is only a trace of atmosphere, and life seems to be out of the question.

Venus is as different from Mercury as it could possibly be. Its great brilliance makes it unmistakable; at its best it is even capable of casting a shadow. In size and mass it is almost a twin of the Earth, but it is a nonidentical twin. The dense atmosphere, made up chiefly of carbon dioxide, hides the surface; the ground pressure is about ninety times that of the Earth's air at sea level; and the clouds contain large quantities of sulphuric acid. These factors make Venus intensely hostile to life. Various probes have flown past it, and several have soft-landed there; the Russians have even obtained two pictures direct from the rock-strewn surface, where the temperature is of the order of 500 degrees C. Venus is unique in having a rotation period longer than its period of revolution around the Sun. To complete the strangeness of the picture, it spins in a wrong-way or retrograde (westward) direction. Like Mercury, it has no satellite and is a solitary wanderer in space.

Earth is in some ways exceptional. In the Solar System, it alone has an atmosphere and temperature suited for advanced life of the kind we can understand, and it alone has appreciable surface water. Moreover, its one natural attendant, the Moon, is oversized for a mere satellite, and there are grounds for regarding the Earth-Moon system as a double planet. As everyone knows, the Moon has been reached; twelve astronauts have walked upon the bleak lunar surface; and many samples of moon rock have been brought home for analysis. The lunar atmosphere is virtually nonexistent, and we may be quite sure that no life has ever appeared there.

Mars, the Red Planet, can be extremely brilliant, outshining every other body in the sky apart from the Sun, the Moon, and Venus, though when at its most remote it resembles nothing more than a bright star. In size and mass it is intermediate between the Earth and the Moon. Its carbon dioxide atmosphere is extremely thin, and the surface pressure is below 10 millibars everywhere, so that no water can exist in the liquid form.

Markings on Mars are easy to see even with small telescopes when the planet is well placed. The white polar caps are made up essentially of ice, with a seasonal coating of solid carbon dioxide. There are dark areas, now known to be nothing more significant than albedo features; and there are the reddish-ocher tracts which have always been called deserts (Lowell compared them with the Painted Desert of Arizona). Before the space age it was widely believed that life existed there, but we know better today. The Mariner probes, the Russian vehicles, and the soft-landing Vikings have shown a fascinating landscape of mountains, valleys, ridges, and craters. There are towering volcanoes, one of which (Olympus Mons) rises to a height of 15.5 miles (25 kilometers), and there is abundant evidence of past water activity. Yet no life has been detected, and most authorities, though not all, have come to the reluctant conclusion that it does not exist there; at best it must be very lowly. The two satellites of Mars, Phobos and Deimos, are irregular lumps of material less than 19 miles (30 kilometers) in diameter, so that they are in no way comparable with our own massive Moon. Future Martian colonists will find them of little use as sources of illumination during the night!

The minor planets, or asteroids, are the junior members of the Sun's family. Only Ceres, Pallas, and Vesta are definitely over 300 miles (500 kilometers) in diameter, and none can retain any trace of atmosphere. It has been suggested that the asteroid swarm is the result of the breakup of a former planet or planets, but it seems more likely that the asteroids never formed part of a larger body. Even if lumped together, they would not make up one world the mass of the Moon.

Probably the most interesting asteroids are those which depart from the main zone. Some, such as Eros, swing closer to the Sun and may bypass the Earth. Icarus, a midget no more than a kilometer in diameter, ventures inside the orbit of Mercury so that at perihelion it must glow red-hot. The Trojans move in the same orbit as Jupiter, though they keep respectfully either well ahead or well behind the

Giant Planet and are in no danger of being swallowed up. Of special significance is Chiron, discovered by Charles Kowal from Palomar in 1977. The diameter may be as much as 400 miles (650 kilometers), which by asteroidal standards is large, but the main interest is in its orbit, which lies mainly between those of Saturn and Uranus. The discovery of Chiron was entirely unexpected, and its nature remains problematical. It could be either an exceptional asteroid, a surviving remnant of the "building blocks" that condensed to form the outer planets; the brightest member of a trans-Saturnian asteroid swarm; or an escaped satellite of Saturn. For the moment it seems best to regard it as asteroidal, but Kowal himself summed up the situation neatly when he commented that Chiron was—well, "just Chiron"!*

Jupiter, the senior member of the Sun's retinue, is a magnificent telescopic object. Its yellowish, flattened disk is streaked with darker belts; there are spots, wisps, and festoons, and usually the Great Red Spot is a prominent feature, though admittedly it sometimes vanishes for a while. Much of our current knowledge comes from *Pioneers 10* and *11*, which bypassed Jupiter in 1973 and 1974 respectively, and the *Voyagers* of 1979. According to the best available information, Jupiter has a relatively small silicate core surrounded by layers of liquid hydrogen which are overlaid by the gaseous surface we can see. The details are always changing, and the surface is in a state of constant turmoil. The Great Red Spot, once thought to be a kind of floating island, is now known to be a whirling storm—a phenomenon of Jovian weather conditions.

Jupiter sends out more energy than it would if it depended entirely upon the Sun, but there is no possibility that it is a small star. The core temperature is far too low to trigger nuclear reactions, and it seems that the extra energy comes from heat left over, so to speak, from the early stages of planetary formation. There is a powerful magnetic field and there are zones of intense radiation which would prove instantly lethal to any astronaut foolish enough to venture inside them. *Voyager 1* established the existence of a thin ring, much too dim to be

---

*Do not confuse *Chiron* with *Charon*, the newly discovered satellite of Pluto. The similarity in names is unfortunate, but there is no direct connection. In mythology, Chiron was the wise centaur who taught many of the great heroes, including Jason and other Argonauts; while Charon was the boatman who ferried new arrivals across the river Styx at the entrance to Pluto's underworld, Hades.

seen with Earth-based telescopes and in no way comparable with the glorious ring system of Saturn.

Jupiter's four large satellites, Io, Europa, Ganymede, and Callisto, are visible with any small telescope and are of planetary size; only Europa is smaller than our Moon, and Ganymede and Callisto are about the same size as Mercury. The Voyager results have shown that Io is of special interest. It has a red surface and active volcanoes. On the other hand, Ganymede and Callisto are "icy" and cratered. Of Jupiter's small satellites, Amalthea moves within the orbit of Io, so that it is in the thick of the Jovian radiation zones; there are at least two more small, close-in satellites, but the others have orbits well beyond that of Callisto and may be captured asteroids rather than bona fide satellites. The outermost four have retrograde motion.

Saturn is the showpiece of the Solar System. Like Jupiter, it has a gaseous surface, and the mean density of its vast globe is actually less than that of water. Belts are visible, as are occasional spots, though none so prominent or so persistent as the Great Red Spot on Jupiter. The rings are made up of rocky or icy particles, each of which moves around Saturn in the manner of a midget moon. They may or may not be due to the breakup of a former satellite that wandered too close to Saturn and was disrupted by the strong gravitational pull.

Whether this is true or not (many astronomers today are inclined to doubt it), Saturn still has over a dozen known attendants. Of these the most important is Titan, which has a diameter of 3600 miles (5800 kilometers) and is therefore intermediate in size between Mercury and Mars. It has an atmosphere with an estimated ground pressure of at least 100 millibars and will be one of the main targets for *Voyager 1* in late 1980. Four more members of the satellite family (Iapetus, Rhea, Tethys, and Dione) are over 600 miles (1000 kilometers) across; the rest are considerably smaller. The outermost member of the system, Phoebe, has retrograde motion and may be classed as asteroidal.

Uranus and Neptune have many points in common, though they are not identical. For instance, it is believed that Neptune has an internal heat source, while Uranus probably does not. They rank as giants, though their diameters are less than half that of Saturn. Recent measurements make Uranus slightly the larger of the two, though it is the less massive (fourteen times the mass of the Earth, as against seventeen for Neptune). Both differ in structure and makeup from either Jupiter or Saturn, and both are too remote to show any real surface detail. Uranus is distinctly greenish in color, Neptune is blue.

Uranus is peculiar inasmuch as it has an axial inclination of 98 degrees. This is more than a right angle, so that technically the rotation is retrograde. The reason for this is unknown. It was once thought that the rotation period was on the order of ten to eleven hours, but it now seems that twenty-three hours is more likely, though we have to admit that there is considerable uncertainty.

All the five satellites move in near-circular orbits in the plane of the planet's equator. Titania and Oberon were discovered by William Herschel in 1787, only six years after he had found Uranus itself. Ariel and Umbriel were added to the list by William Lassell in 1851, while Miranda was discovered by G. P. Kuiper in 1948. All are rather faint, and all are smaller than our Moon; their diameters range from about 1120 miles (1800 kilometers) (Titania) down to about 340 miles (550 kilometers) (Miranda). It cannot be said that there is anything remarkable about the satellite system.

The rings of Uranus came to light in an unexpected manner. One way to measure the diameter of a remote object is to note the time taken for it to pass over and occult a star. This method, pioneered by Gordon Taylor of the Royal Greenwich Observatory in England, gives excellent results but means waiting for a suitable occultation. A good chance came on 10 March 1977, when Taylor calculated that Uranus would pass in front of a ninth-magnitude star known by its catalogue number SAO 158687. The results were surprising. Before the actual occultation, and again afterwards, the star gave a series of "winks" which could be due only to its being briefly covered by a set of rings associated with Uranus. Confirmatory observations have since been obtained. The rings are not like those of Saturn; they are dark instead of being highly reflective, and they are relatively narrow. (They have no connection with the Uranian rings once reported by Herschel, which do not exist; they were optical effects. Even Herschel could not be right all the time!)

From Uranus, sunlight would still be fairly strong, more than a thousand times brighter than full moonlight on Earth, but the planets would never be seen to advantage. Saturn would be prominent enough when well placed, every forty-five years or so, but Jupiter would remain inconveniently close to the Sun in the sky, and the inner planets would be, to all intents and purposes, out of view. Neptune would be just visible with the naked eye when near opposition, but not Pluto.

Neptune does not share Uranus's unusual axial tilt, but the satellites are of great interest. Of Neptune's two known attendants, Triton is large—perhaps even larger than Titan in Saturn's family—and has an almost circular orbit; but it has retrograde motion, in which it is unique among nonasteroidal satellites. Nereid, no more than 300 miles (500 kilometers) in diameter, has direct motion, but a path so eccentric that it resembles that of a comet. All in all, there are indications that strange events have taken place in the outer reaches of the Solar System.

Leaving the planets and their satellites for the moment, let us complete this hasty survey of the solar family by noting the less substantial members—the comets, meteoroids, and interplanetary particles.

A comet has been described as "the nearest approach to nothing which can still be anything." According to a theory developed by F. L. Whipple, a typical comet has a nucleus made up of a conglomerate of rocky fragments held together with ices such as frozen methane, ammonia, carbon dioxide, and water. When the comet approaches the Sun, these substances begin to evaporate so that a tail develops—only to vanish again as the comet recedes once more into the depths of space.

Many small comets have eccentric orbits which carry them around the Sun in periods of a few years, but the brilliant comets seen now and then throughout history have much longer periods, so that they cannot be predicted. The only exception is Halley's comet, which returns every seventy-six years and is due back once more in 1986. A comet is so flimsy, judged by planetary standards, that it suffers violent perturbations in its orbit whenever it encounters a massive body such as Jupiter, and in some cases the orbit may be completely changed.

Meteors, or shooting stars, are often classed as cometary debris. They are tiny, friable bodies which burn away when they dash into the Earth's upper air. Meteorites are of different nature; they may land intact. Some of them are really massive. The present holder of the heavyweight record is the Hoba West Meteorite, which is still lying where it fell in prehistoric times, in Southern Africa. Nobody is likely to run away with it, since its weight is at least 60 tons. And, of course, the Meteor Crater in Arizona, over 3900 feet (1200 meters) across, was certainly produced by a meteoritic fall which took place before there were any astronomers to observe it. Fortunately, such major falls are rare, and there is no reliable record of any human fatality due to a plunging meteorite. Finally, the Solar System contains a great deal of thinly spread interplanetary material that betrays its presence when lit up by the Sun to produce the cone-shaped glow we call the Zodiacal Light.

This, then, is a picture of the Solar System as we know it today. But to begin the real story of Pluto we must go back to the eighteenth century, when Saturn was still believed to be the outermost of the planets.

3

The Discovery of
Uranus, 1781

by Patrick Moore

Friedrich Wilhelm Herschel, always remembered today as William Herschel, was born in 1738. His father was a musician in the Hanoverian Army, and William began his career in the same way. At the age of fourteen he joined the army, and, in 1757, during the Seven Years' War, he even came under fire at the Battle of Hastenbeck. However, a military life—even as a musician—was not to his liking, so he obtained his discharge and decided to go to England. For nine years he held various musical posts and then, in 1766, became organist at the fashionable Octagon Chapel in Bath. When he next visited his family home in Hanover, his father had died, and his sister Caroline was unhappy. Caroline came to England with William and became her brother's devoted assistant. Rather surprisingly, William did not become a naturalized Englishman until he was in his mid-fifties.

William Herschel had always been interested in astronomy, but it was not until he had established himself as a successful organist that he was able to devote much time to what was then nothing more than a casual hobby. Caroline related that "he used to retire to bed with a bason [*sic*] of milk or glass of water, and Smith's *Harmomics* [*sic*] and *Optics*, Ferguson's *Astronomy*, &c., and so went to sleep buried under his favourite authors; and his first thoughts on rising were how to obtain instruments for viewing those objects himself of which he had been reading." Telescopes were expensive items so Herschel decided to make reflectors for himself. Of course he had his failures, but in the end he became the best telescope-maker of his day. His largest reflector, with a focal length of 40 feet (12.2 meters) and an aperture of 49 inches (124.5 centimeters), was not surpassed in size until the Earl of Rosse built the "Leviathan of Parsonstown," a 72-inch (182.9-centimeter) one, in 1845.

In fact, most of Herschel's important work was done with much smaller telescopes, and the reflector with which he made his epic discovery in 1781 had an aperture of only 6.2 inches (15.7 centimeters). Mirrors in those days were made not of glass, but of speculum metal, an alloy of copper and tin. Herschel even cast his own blanks. (One of his first molds was made of horse dung, but it sprang a leak at the wrong moment, with disastrous results!)

Herschel was interested chiefly in the stars, and he was anxious to find out the way in which the stars of our galaxy were arranged. From August 1779 he busied himself with a "review of the heavens," taking in stars down to the eighth magnitude. On 13 March 1781, using his 6.2-inch (15.7-centimeter) reflector, he came across something distinctly unusual. In his own words:

> While I was examining the small stars in the neighbourhood of H Geminorum, I perceived one that appeared visibly larger than the rest; being struck with the uncommon magnitude, I compared it to H Geminorum and the small star in the quartile between Auriga and Gemini, and finding it so much larger than either of them, suspected it to be a comet . . . . The power I had on when I first saw the comet was 227. From experience I knew that the diameters of the fixed stars are not proportionally magnified with higher powers, as the planets are; therefore I now put on the powers of 460 and 932, and found the diameter of the comet increased in proportion to the power, as it ought to be, on a supposition of its not being a fixed star, while the diameters of the stars to which I compared it were not increased in the same ratio.

This account makes it quite clear that Herschel identified the ob-

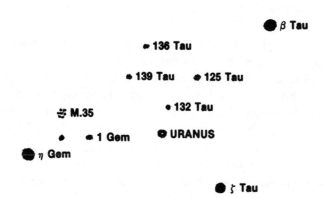

Discovery position of Uranus, 13 March 1781.

ject because of its appearance—eloquent testimony to his eyesight as well as to the quality of his telescope. Over the next few nights he observed definite motion by "about 2¼ seconds per hour," and he sent a paper to the Royal Society in London, headed "An Account of a Comet." The position which he gave was pleasingly accurate, and both of his comparison stars can be identified. H Geminorum is now known as 1 Geminorum and is of magnitude 4.3; the "small star in the quartile" is presumably 132 Tauri, just below magnitude 5.

Herschel's paper was read to the Royal Society on 26 April. By then the new body had been observed by several astronomers, including Charles Messier, the famous French comet-hunter, and the English Astronomer Royal, Nevil Maskelyne. Messier identified the object because of its movement, but not by its appearance, and went so far as to write to Herschel: "Nothing was more difficult than to recognize it; and I cannot conceive how you have been able to return several times to this star or comet; for it has been necessary to observe it for several consecutive days to perceive that it was in motion."

Meanwhile, Maskelyne had begun to suspect that Herschel's discovery was more important than it appeared at first sight, and in a letter to Herschel written three days before the Royal Society meeting, he wrote: "I am to acknowledge my obligation to you for the communication of your discovery of the present comet, or planet, I don't know which to call it. It is as likely to be a regular planet moving in an orbit nearly circular round the sun as a Comet moving in a very eccentric ellipsis."

Charles Messier had the same idea and on 29 April recorded that the object "has none of the distinctive character of comets, as it does not resemble any of those I have observed." But everything depended upon the orbit and the distance. On 8 May the French mathematician Bouchart de Saron—later guillotined during the Revolution—announced that the object was very remote and must lie well beyond the orbit of Saturn, in which case it could only be a new planet. Apparently the first reasonably accurate orbit was worked out by the Finnish astronomer Anders Lexell, then working at St. Petersburg, who gave a period of between 82 and 83 years, and a mean distance from the Sun of 18.9 astronomical units.* As the true mean distance of Uranus is 19.2 astronomical units, and the period is 84 years, Lexell was very close to the mark. He was also right in maintaining that the new world was larger than any of the other planets apart from Jupiter and Saturn.

It is natural to assume that Herschel was elated. Though he had not been deliberately searching for planets, he had made the most of his opportunity, and as he said in a letter to a friend of his, Dr. Hutton: "Had business prevented me that evening, I must have found it the next, and the goodness of my telescope was such that I must have perceived its visible planetary disk as soon as I looked at it."

Obviously, a name had to be selected. By the end of 1782 Herschel had received royal recognition from King George III of England and Hanover, and in gratitude he proposed to call his planet "Georgium Sidus," the Georgian Star. It is hardly surprising that this found no favor with foreign astronomers, quite apart from the fact that the object was not a star! Jean Bernoulli, of Berlin, suggested "Hypercronius" (that is, "above Saturn"); from Uppsala in Sweden, Erik Prosperin proposed "Neptune"; the French astronomer Lalande, one of the earliest observers of the planet, wanted to call it simply "Herschel." Johann Elert Bode, of Germany, preferred "Uranus," in honor of the first ruler of Olympus—Saturn's father. Other mythological names were also put forward, but Uranus was clearly the favorite, and eventually it prevailed. Yet the final acceptance was delayed for a surprisingly

---

* One astronomical unit is defined as the distance between the Earth and the Sun: approximately 93,000,000 miles (150,000,000 kilometers).

long time. Up until 1850 the influential *Nautical Almanac*, which had been founded by Maskelyne, still used the name of "the Georgian."

Herschel's reputation was made. Henceforth he could devote his whole time to astronomy, with music taking a very secondary place, and before his death in 1822 he had been awarded virtually every honor that the scientific world could bestow. As an observer he has had few, if any, peers; to list all his discoveries would take many pages and would be beyond the scope of this book. Yet some of his ideas sound strange today. He regarded the habitability of the Moon as "an absolute certainty," and he was even convinced that life existed in a cool, pleasant region below the surface of the Sun!

Though Herschel was the first to recognize Uranus as being non-stellar, he was not the first to observe it. Diligent searches through past observations showed that the planet had been recorded over twenty times between 1690 and 1771. The Rev. John Flamsteed, the first Astronomer Royal at Greenwich, saw it on 23 December 1690, and even gave it a number—34 Tauri. He observed it again in 1712, and for a third time in 1715, still without realizing that it was anything but a star. John Bradley, the third Astronomer Royal, recorded it in 1748 and 1750, and Johann Tobias Mayer, of Germany, in 1756. However, the most remarkable "run" was due to Pierre Charles Lemonnier, of Paris. Altogether he made a dozen observations—eight of them over a period of a month, from 27 December 1768 to 23 January 1769. Had he been blessed with an orderly mind he could not have failed to make the discovery, but he failed to compare his observations, one of which was eventually found scrawled upon a paper bag which had once contained hair perfume. (Lemonnier must indeed have been a difficult colleague. It was said that he never failed to quarrel with anyone he met.)

These prediscovery observations proved to be of great value later on, because they showed up the discrepancies between Uranus's calculated position and its actual position in the sky. Meanwhile, attempts were made to measure the size of Uranus, which was no easy matter, since the apparent diameter is never as great as 4 seconds of arc. Herschel, in 1788, gave a value for the real diameter of 34,217 miles (55,100 kilometers), which is of the right order even though it may be slightly too great. The value generally accepted today is 31,981.5 miles (51,500 kilometers), though it is worth noting that in 1977 J. Janesick,

Statue of Flamsteed at Herstmonceux. (*Photo by Patrick Moore*)

of the Mount Lemmon Observatory in the United States, gave a value of 34,155 miles (55,000 kilometers), in excellent agreement with Herschel's determination so long before.

The discovery of Uranus was a triumph of observation. It more than doubled the size of the known Solar System; remember, the mean distance of Uranus from the Sun is more than twice that of Saturn. Even so, the system was not complete. Within twenty years of Herschel's triumph, another new chapter in planetary discovery opened.

4

Bode's Law and the Asteroids

by Patrick Moore

In 1800 six astronomers met at the Observatory of Lilienthal, a private establishment owned and run by the chief magistrate of the town, Johann Hieronymus Schröter. The six had a major project in view, which they expected would keep them occupied for many years. They deliberately set out to hunt for a new planet moving around the Sun between the orbits of Mars and Jupiter.

At first sight this might not seem to be a really profitable exercise. Any new planet lying in this part of the Solar System would necessarily be small; otherwise it would have come to light long ago. All the same, Schröter and his colleagues had good reason to suspect that it might exist. They based their hopes on a curious relationship known as Bode's Law.

In fact, this is an unfortunate name. The law is not a true law; as we will see, it breaks down hopelessly in the outermost parts of the Solar System, and in any case it is inexact. Neither was it discovered by Johann Elert Bode (the same Bode who proposed the name of Uranus

for Herschel's planet). The first mention of the so-called law was made
in 1766 by an otherwise obscure professor named Johann Daniel Titius,
of Wittenberg. He had translated a book (*Contemplation de la Nature*)
by the French philosopher Charles Bonnet and had inserted some
material of his own. In chapter 4, between paragraphs 6 and 8 of the
translation, he added the following note:

> For once pay attention to the widths of the planets from each
> other and notice that they are distant from each other almost in
> proportion as their bodily heights increase. Given the distance
> from the Sun to Saturn as 100 units, then Mercury is distant 4
> such units from the Sun; Venus, 4 + 3 = 7 of the same; the Earth
> 4 + 6 = 10; Mars 4 + 12 = 16 etc. But see, from Mars to Jupiter
> there comes forth a departure from this so exact progression. From
> Mars follows a place of 4 + 24 = 28 such units, where at present
> neither a chief nor a neighbouring planet is to be seen. And shall
> the Builder have left this place empty? Never!

At first little notice was taken of this interesting little comment.
Then, in 1772, Johann Elert Bode came across it and was impressed.
He republished it—without, at first, giving any credit to Titius—and
modified it to some extent. Titius had suggested that there might be
undiscovered satellites in the gap between the orbits of Mars and
Jupiter, but this idea was out of court at once, for obvious reasons.
Bode preferred the theory that a new principal planet lay in the gap
corresponding to number 28.

There is no doubt that Titius was the first to announce the relationship, and in all fairness it should be named after him. However, it
was Bode who made it famous, and as it is always called Bode's Law, I
propose to follow convention.

Let us sum up what is meant. Take the numbers 0, 3, 6, 12, 24,
48, 96, and 192, each of which (apart from the first) is double its
predecessor. Add 4 to each. Taking the Earth's distance from the Sun
as 10, the remaining figures give the mean distances of the planets
with fair accuracy out as far as Saturn, the outermost planet known in
1772.*

---

* One may convert the Bode units to astronomical units by dividing by 10, that is, shifting
the decimal point to the left.

	Distance from the Sun	
Planet	By Bode's Law	Actual
Mercury	4	3.9
Venus	7	7.2
Earth	10	10
Mars	16	15.2
. . . .	28	. . . . .
Jupiter	52	52.0
Saturn	100	95.4

In 1781 Herschel discovered Uranus. On the Bode scale its distance should have been 192 + 4 = 196. Actually it proved to be 191.8, which was not far wrong, and most astronomers came to the conclusion that Bode's Law was of real significance.

There remained the problem of the missing body in the position corresponding to Bode's 28. Even Johannes Kepler, in the early seventeenth century, had commented that "between Mars and Jupiter I put a planet." Since any new member of the Solar System would have to be below the limit of naked-eye visibility, a systematic hunt would have to be made, and this was the aim of the astronomers who met at Lilienthal in 1800.

Johann Schröter was a keen observer and a great enthusiast. He owned a telescope which had been made for him by Herschel, as well as a larger reflector by Schräder of Kiel; he concentrated upon observations of the Moon and planets and may be regarded as the real father of selenography (the study of the lunar surface). His somewhat clumsy draftsmanship has led to his being generally underrated, but an examination of his work shows that he accomplished a great deal. *
Also present at the meeting was the Hungarian baron Franz Xaver von Zach; another delegate was a German doctor and amateur astronomer, Heinrich Olbers. The other three members were Karl Harding (afterwards Schröter's assistant), Gildemeister, and Von Ende.

The six nicknamed themselves the Celestial Police and decided to divide up the Zodiac into a number of zones, each of which was to be patrolled by one or more astronomers. This would involve persuading

---

* Schröter's observatory was destroyed by the invading French soldiers in 1814, and his telescopes were plundered because the brass tubes were mistaken for gold. Unhappily, many of Schröter's original observations were also lost, which was a tragedy for science. Schröter was too old to begin again; he died two years later.

others to join in the search, and messages were sent out. One of them went to Giuseppi Piazzi, director of the Palermo Observatory in Sicily, but by an ironical twist of fate Piazzi had already made the key discovery before he had received any word from Lilienthal. He had been compiling a star catalogue, and on the night of 1 January 1801—the first day of the new century—he noted an eighth-magnitude object in Taurus with which he was not familiar. On the next night he found that it had moved. Therefore, it could only be a member of the Solar System.

Piazzi was suitably cautious and announced that he had discovered something which was probably a comet, though in a letter to his friend B. Oriani he dropped a hint that it could be a planet. When von Zach heard of the discovery, he immediately assumed that Piazzi's object was the long-sought planet between the orbits of Mars and Jupiter. The trouble was that by now the object had vanished into the evening twilight, and there was no chance of observing it again before the early autumn of 1801. Neither had Piazzi been able to continue his observations for long enough for a reliable orbit to be computed, and for a time it was thought that the object had been totally and permanently lost.

From August 1801 many observers set out to relocate it, using orbits calculated by various mathematicians including Olbers and von Zach. They failed. Luckily a brilliant young German named Karl Gauss came to the rescue. He had read about the supposed new planet, and in his own words:

> Nowhere in all the annals of astronomy do we find such an important occasion; and scarcely is it possible to imagine a more important opportunity for pointing out, as emphatically as possible, the importance of that problem, as at the moment when every hope of re-discovering, among the innumerable little stars of heaven, that mite of a planet which had been lost to sight for nearly a year, depended entirely upon an approximate knowledge of its orbit, which must be deduced from these scanty observations.

Gauss took up the challenge. He developed an entirely new mathematical method, worked out where the object should be, and sent his results to von Zach. On December 31, von Zach identified the object in almost precisely the position given by Gauss. Olbers found it independently on the following night, exactly a year after Piazzi's original discovery.

There could no longer be any doubt that the object was planetary rather than cometary. Gauss gave the distance as 27.7 on the Bode scale, which corresponded very well with the expected 28. Piazzi was given the honor of choosing a name; he selected that of Ceres, the patron goddess of Sicily.

As expected, Ceres was very small; we now know the diameter to be just over 600 miles (1000 kilometers). It came as no real surprise when, on 28 March 1802, Olbers discovered a second dwarf planet, now named Pallas, in the same region of the Solar System. This time the distance on the Bode scale was 26.7, but the orbit was rather more eccentric than that of Ceres and was inclined at the sharp angle of 35 degrees. Olbers made the bold suggestion that Ceres and Pallas had once been parts of a single body which had met with disaster. In this case other small planets might exist, and the Celestial Police continued their work. On 2 September 1804, Harding, at Lilienthal, discovered minor planet No. 3, Juno, and on 28 March 1807, the indefatigable Olbers detected No. 4, Vesta. No more seemed to be forthcoming, and the Police finally ceased their activities in 1815.

Years later the problem was taken up by a lone amateur, Karl Ludwig Hencke, postmaster of the little German town of Driessen. Following the same procedure as the Celestial Police—though without any assistance—he searched away, and at last, on 8 December 1845, his patience was rewarded; he discovered minor planet No. 5, now called Astræa. Since then thousands of small worlds have been found, though most of them are very small and faint.

At the beginning of 1846, the known planetary bodies were twelve in number: the main planets (Mercury to Uranus) and the five asteroids. Already there were strong indications of another world moving well beyond the path of Uranus, and on Bode's scale its distance would be 388 units. The stage was set for another spectacular leap forward in the exploration of the Solar System.

5

The Search for Neptune

by Patrick Moore

As we have seen, the discovery of Uranus gave new support to the view that Bode's Law was something more than mere coincidence. When the first asteroids were discovered between 1801 and 1807, there no longer seemed any doubt. What, then, about a new planet beyond Uranus, at a distance corresponding to 388 units on Bode's scale? Even if the planet were large, it would be well below naked-eye visibility, and at first there was no reason to suppose that it existed at all.

A hint of approaching problems came when the orbit of Uranus was computed. Remember, there were some prediscovery observations, going back to Flamsteed's record of 1690, and if these were accurate they would be of great value. In 1788 a German astronomer, Placidus Fixlmillner of Kremsmunster, checked on the old observations and found that those of 1690 (by Flamsteed) and 1756 (by Mayer) could not be reconciled with the postdiscovery observations made since 1781. Fixlmillner could give no solution, and there, for a while, the matter rested.

Thirty years later Alexis Bouvard, already famous as a mathematician, made a full investigation of the movements of Uranus based upon tables worked out by his fellow countryman Delambre. Once again he ran into trouble. The old observations could not be made to fit in with the new ones. Bouvard did the only possible thing; he rejected them and started again. After making new calculations based entirely upon the results obtained since 1781, he felt confident that all would be well.

Yet was it wise to dismiss the old observations so summarily? Even though Uranus had not been recognized as a planet, men such as Flamsteed, Bradley, and Mayer (or even Lemonnier!) were not likely to be guilty of gross errors. One man who was not convinced was Friedrich Wilhelm Bessel, later to become famous for making the first measurement of the distance of a star. Bessel believed that there had to be some other explanation, and he began to make a fresh series of observations. Sure enough, discrepancies began to creep in once more, and less than four years after the publication of Bouvard's results it had become painfully clear that something was wrong.

So far as is known, the first man to make a definite suggestion about the existence of a trans-Uranian planet was an amateur astronomer: the Rev. T. J. Hussey, Rector of Hayes in Kent. He was a good observer and was equipped with an adequate telescope. On 17 November 1834, he wrote to George Biddell Airy, then Plumian Professor of Astronomy at Cambridge, and from 1835, Astronomer Royal in succession to John Pond.* Hussey worded his letter as follows:

> Having taken great pains last year with some observations of Uranus, I was led to examine Bouvard's tables of that planet. The apparently inexplicable discrepancies between the ancient and modern observations suggested to me the possibility of some disturbing body beyond Uranus, not taken into account because unknown. My first idea was to ascertain some approximate place of the supposed body empirically, and then with my large reflector set to work to examine all the minute stars thereabouts; but I found myself totally inadequate to the former part of the task. . . . I may be wrong, but I am disposed to think, such is the perfection of my equatoreal's object-glass, I could distinguish, almost at

---

*Pond was the only Astronomer Royal who was asked to resign because he neglected his duties. In all fairness it was not entirely his fault, because he suffered from ill health. He left Greenwich Observatory in such a poor state, that Airy had to reestablish its reputation, which he did with great energy and skill.

once, the difference of light of a small planet and a star. My plan of proceeding, however, would be very different. I should accurately map the whole space within the required limits, down to the minutest star I could discern; the interval of a single week would thus enable me to ascertain any change.

Several important points are raised in this letter. First, Hussey had become convinced that there was some disturbing body beyond Uranus—and since Uranus is a giant, the unknown planet also would have to be large, as otherwise it would have no measurable effects. Secondly, Hussey realized that the planet would be likely to show a disk, in which case it would be comparatively easy to recognize. And thirdly, he was sensible enough to appreciate that the actual calculation of the position of the hypothetical planet was beyond his mathematical powers. It is only reasonable to assume that he wanted Airy's help.

Unfortunately Airy was skeptical, and his reply was most discouraging. On 23 November he wrote: "I have often thought of the irregularity of Uranus, and since the receipt of your letter have looked more carefully into it. It is a puzzling subject, but I give it as my opinion, without hesitation, that it is not yet in such a state as to give the smallest hope of making out the nature of any external action on the planet." Three years later he wrote to Bouvard: "If it be the effect of any unseen body, it will be nearly impossible ever to find out its place."

Because Airy comes so much into the story, and because he has been so widely blamed for the later series of mishaps, it seems necessary to say a little more about him. There is no doubt that he was a brilliant scholar, and he accomplished a great deal of valuable work. Neither is there any doubt that he was very much aware of his own importance, and he had certain "blind spots" in his mental makeup. He also had a passion for order which amounted to an obsession; it is related that on one occasion he spent a whole afternoon writing *Empty* on large cards to be nailed onto empty packing boxes. He also insisted that his observers should remain on duty at night even when rain was falling, and according to one account he would "do the rounds" late in the evening, visiting the various domes at Greenwich Observatory and saying "Mr. So-and-so. You *are* there, aren't you?" In replying to Hussey then, Airy showed that in his view any attempt to locate a trans-Uranian planet would be abortive. Not until too late did he change his mind.

However, as the wanderings of Uranus continued, astronomers everywhere started to come around to Hussey's theory (though there is no record that Hussey himself took any further part; no doubt he accepted Airy's decision as final). Various ideas were bandied about. It was suggested that there might be a resisting medium in the outer part of the Solar System, though this seemed improbable for many reasons. A large perturbing satellite of Uranus was also proposed, or even a past collision with a comet which knocked Uranus off its course. However, the "unknown planet" idea seemed to be by far the most logical. It was supported in 1835 by Friedrich Nicolai, of Mannheim, who pointed out that there were perturbations in the movement of Halley's comet which could not be explained, and that perihelion passage in 1835 had been a day late. Nicolai wrote, "One immediately suspects that a trans-Uranian planet (at a radial distance of 38 astronomical units, according to the well-known rule), might be responsible for this phenomenon." Again, note the almost religious belief in Bode's Law.

Meanwhile Bessel, now director of the Observatory of Königsberg, was paying close attention to the problem of Uranus. He was highly skeptical of Bouvard's claim that the prediscovery observations were unuseable, and in a public lecture delivered in 1840 he made his views known. Unlike Airy, he believed that by studying the motions of Uranus it would be possible to track down the unknown body which was causing the trouble, and together with his pupil, F. W. Flemming, he prepared to begin work. Unhappily, Flemming died suddenly and unexpectedly; Bessel became ill and was never able to follow the problem up. Ironically, he died six months before the new planet was found.

The next step was taken by a young student, John Couch Adams, who had been born at Lidcot in Cornwall in 1819 and had shown early signs of exceptional mathematical ability. He entered St. John's College, Cambridge, and became interested in problems of celestial mechanics. A memorandum dated 3 July 1841 reads:

> Formed a design, at the beginning of this week, of investigating, as soon as possible after taking my degree, the irregularities in the motion of Uranus, which were as yet unaccounted for; in order to find whether they may be attributed to the action of any undiscovered planet beyond it, and if possible thence to determine the elements of its orbit, &c., approximately, which would probably lead to its discovery."

To a fellow student, George Drew, he confided, "Uranus is a long way out of his course. I mean to find out why. I think I know."

Adams passed his degree—brilliantly—and then set to work. A few months of research convinced him that the project was feasible. As early as 1843 he had some preliminary results, but obviously he needed the latest data, and so he contacted the Cambridge professor of astronomy, James Challis.

Challis was of very different caliber from Airy. Far from being a dominating character, he was hesitant and ultraconventional. However, he contacted Airy, mentioned that Adams was working on the problem of Uranus, and asked for the latest tables of the planet. On 15 February 1844 Airy duly sent them, and Challis passed them on to Adams. By September 1845 it seemed that the problem had been solved. Adams had found out where the new planet ought to be.

Obviously he went to see Challis, and equally obviously Challis suggested sending the results to Airy at Greenwich. Challis even wrote a letter of introduction.

> My friend Mr. Adams (who will probably deliver this note to you) has completed his calculations regarding the perturbation of the orbit of Uranus by a supposed ulterior planet, and has arrived at results which he would be glad to communicate to you personally, if you could spare him a few moments of your valuable time. His calculations are founded on the observations you were so good as to furnish him with some time ago; and from his character as a mathematician, and his practice in calculation, I should consider the deductions from his premises to be made in a trustworthy manner.

This was the real beginning of the chapter of accidents which followed. Adams went to Greenwich, only to find that Airy was abroad attending a conference. Airy came back a few days later and wrote to Challis that he was "very much interested in the subject of his investigations," and that he would be delighted to hear more about them. On 21 October 1845 Adams again called at the Observatory. When he arrived, Airy was out. Adams left a card, which was taken to Mrs. Airy and then forgotten. When Adams returned, the butler told him that the Astronomer Royal was having dinner and could not be disturbed. This was not Airy's fault (evidently he was not even told that there had been a visitor), but Adams felt rebuffed. All he could do was to leave his paper for Airy to see.

Remember, Airy was still far from satisfied that the problem was capable of solution; he was not tolerant of youth, and Adams was a mere stripling in his midtwenties. After a fortnight Airy sent a reply, asking a question about the results which showed, only too clearly, that he had not really appreciated the nature of the problem. Adams did not answer because, as he said later, he regarded the question as too trivial to be worthy of comment. It was characteristic of Airy that he made no attempt to follow the matter up. He felt that "Adams' silence . . . was so far unfortunate that it interposed an effectual barrier to all further communication. It was clearly impossible for me to write to him again."

But if Airy was unenthusiastic, then so was Challis. Adams had handed him what he believed (rightly) to be a solution, and he also believed (again rightly) that his hypothetical planet would show a small disk, so that it could not be confused with a star. This was how Herschel had recognized Uranus, and Hussey had had the same idea. At Cambridge there was the 11.75-inch (29.85-centimeter) Northumberland equatorial (so called because it had been paid for by the Duke of Northumberland), whereas at Greenwich the largest telescope was the modest 6.7-inch (17.02-centimeter) Sheepshanks refractor. Had Challis been really interested, why did he not undertake a preliminary search? But Challis was not that kind of man; he was never ready to depart from strict, conventional procedure. He had his set program, and he was not prepared to undertake anything novel.

There was also another unfortunate episode. In October 1845 Airy showed Adams's results to the Rev. William Rutter Dawes and asked for his views. Dawes, a well-known amateur astronomer, was enthusiastic and wrote to his friend William Lassell, who lived in Liverpool and had built himself a fine 24-inch (61-centimeter) reflector. Lassell was in bed with a sprained ankle; before he had recovered, the maid had destroyed Dawes' letter, and again nothing was done.

Meanwhile, things had been happening abroad. The leading French astronomer of the time was François Arago, director of the Paris Observatory and secretary to the Academy of Sciences. Arago's attention had been taken by the skill of a young chemist-turned-astronomer, Urbain Jean Joseph Leverrier, who had started to make himself a great reputation in celestial mechanics. During the summer of 1845 Arago decided that Leverrier was just the man to tackle the problem of Uranus. In June of that year Leverrier started work—quite unaware that Adams, in England, was already well on the way toward the solution.

Leverrier wasted no time, and before the end of the year he had published his first treatise on the subject. Airy read it in December and described it as "a new and most important investigation." Yet still he took no action; he remained unwilling to order a search to be made, and neither did he tell Leverrier anything about Adams's work.

On 1 June 1846 Leverrier produced his second memoir, *Recherches sur les Mouvements d'Uranus*. He had become firmly convinced that the unknown planet must be more remote than Uranus (otherwise, it would have exerted marked perturbations upon Saturn), and he was also, consciously or not, influenced by Bode's Law. He gave a position for the planet which he felt confident would not be far from the truth. But just as Adams had had no help from practical observers in England, so Leverrier was equally unlucky in France. Nobody began to search—not even Arago. Ironically, the man stirred to action was George Biddell Airy!

Airy read Leverrier's treatise on 23 June, and commented:

> I cannot sufficiently express the feeling of delight and satisfaction which I received from it. The place which it assigned to the disturbing planet was the same, to one degree, as that given by Mr. Adams' calculations, which I had perused seven months earlier. To this time I had considered that there was still room for doubt of the accuracy of Mr. Adams' investigations. . . . But now I felt no doubt of the accuracy of both calculations.

He wrote to Leverrier, asking the same question as he had put to Adams. Leverrier replied promptly and concisely. Yet—and this is a significant point—in his letter to Leverrier, Airy made no mention of Adams. He also discussed the problem with Peter Hansen, director of the Seeberg Observatory in Denmark, who had been on a visit to Greenwich, again mentioning Leverrier but not Adams.

By now Airy was under some pressure. John Herschel, son of the discoverer of Uranus, was anxious for a search to be started, and at last Airy took action. On 9 July 1846 he wrote to Challis. Airy's first sentence sounds strange in view of his past inertia.

> You know that I attach importance to the examination of

that part of the heavens in which there is . . . reason for suspect-
ing the existence of a planet exterior to Uranus. I have thought
about the way of making such examination, but I am convinced
that (for various reasons, of declination, latitude of place, feeble-
ness of light, and regularity of superintendence) there is no prospect whatever of its being made with any chance of success, except
with the Northumberland Telescope.

Challis was away from Cambridge when Airy's letter arrived, and
Airy followed it up on 13 July with a more urgent message:

I have drawn up the enclosed paper, in order to give you a notion
of the extent of work incidental to a sweep for the possible planet.
I only add at present that, in my opinion, the importance of this
inquiry exceeds that of any current work, which is of such a
nature as not to be totally lost by delay.

As soon as Challis returned on 18 July, he replied to Airy's letter,
assuring him that he would undertake the search personally. One
would have assumed that as a practical observer, he would have been
able to organize the hunt without any help. In fact he followed Airy's
advice and adopted a cumbersome procedure which could hold out little
hope of quick success. He concentrated upon a wide band in the sky,
stretching well away from the calculated position of the hypothetical
planet, and set out to map all the stars down to the eleventh magnitude.
This meant checking on well over three thousand stars, and Challis
was not particularly energetic in any case. On 2 September he told
Airy, "I get over the ground very slowly . . . to scrutinize thoroughly in
this way the proposed portion of the heavens will require many more
observations than I can make this year." He did not even check his
results promptly. He merely plodded on, apparently waiting for something to turn up. Moreover, he took time off to observe some faint
comets.

Across the Channel, Leverrier was completing his third memoir,
which he presented to the Academy of Sciences on 31 August. According to him, the planet had a period of 217.387 years, and was at a distance of 36.1 astronomical units from the Sun. Still no search was in-
stituted, and Leverrier lost patience with his countrymen.* He wrote

---

*Incidentally, it has been said that Leverrier was one of the rudest men who has ever
lived. He was forced to resign as director of the Paris Observatory because of his ill
temper, though he was reinstated when his successor, Delaunay, was drowned in a boat-
ing accident. One colleague commented that even if Leverrier was not the most detestable
man in France, he was certainly the most detested!

to Johann Galle at the Berlin Observatory. As soon as Galle received the letter on 23 September, he asked the permission of the observatory's director, Johann Encke, to begin a hunt and Encke agreed. During the discussion they were joined by a student, Heinrich d'Arrest, who asked to be allowed to take part. Galle felt that it would have been "unkind to refuse the wish of this zealous young astronomer." As soon as darkness fell the two men set to work, using a fine 9-inch (22.9-centimeter) Fraunhofer refractor. It must have been a tense time. Unlike Challis, Galle concentrated on the exact position which Leverrier had given—right ascension 22^h [*sic.*]** 46^m, declination −13°24′—hoping to find an object which showed a definite disk. For some time he had no success. D'Arrest suggested using a map and produced the relevant section of a new star chart which had been completed by Bremiker of the Berlin Academy. Galle returned to the telescope, and called out the appearances and positions of the stars while d'Arrest checked at the map. Before long Galle described a star of magnitude 8 at right ascension 22^h 53^m 26^s [*sic*]. "That star is not on the map," exclaimed d'Arrest. The long search was over at last.

Encke joined his two observers in the dome, and the object was followed until it set, but not until the following night were they completely satisfied that the object was indeed a new planet. Encke measured its apparent diameter as 3.2 seconds of arc; Leverrier had given an estimate of 3.3 seconds. Moreover, Leverrier's calculated position for the planet was within one degree of the truth. On 25 September 1846 Galle wrote to Leverrier:

> The planet whose position you have pointed out actually exists. The same day that I received your letter, I found a star of the eighth magnitude which was not shown on the excellent chart (drawn by Dr. Bremiker), Hora XXI of the series of celestial maps published by the Royal Academy of Berlin. The observations made the following day determined that this was the sought-for planet.

Encke followed it up with a letter congratulating Leverrier on his "brilliant discovery."

Challis, still laboring away at Cambridge, saw Leverrier's last

---

** Right ascension 22^h 46^m is the value given in Grossers book, *The Discovery of Neptune.* Obviously this is a serious error. Only the value of 21^h 46^m could fit the 323 degrees of longitude and declination −13°21′.

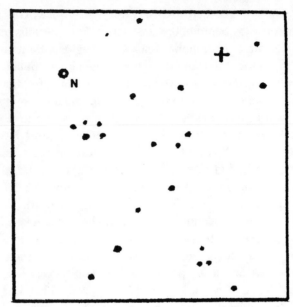

One-degree square of the chart in the region of Neptune's discovery position. N indicates where Neptune actually was, and the cross shows where Leverrier had expected it to be.

memoir on 29 September. Leverrier had maintained that the planet would show a definite disk, and Challis, possibly remembering that Adams had suggested the same thing, began to sweep over the relevant area, looking for anything which seemed to be unstellar. He reported to his assistant that one object looked as though it might show a disk but did not bother to change to a higher magnification; he merely commented that he would do so on the following night. There follows an episode which is probably true, though there is no definite proof. Apparently the next night was clear; the Rev. William Kingsley, of Sidney Sussex College, was dining with Challis when the subject of the supposed planet came up. Kingsley suggested looking again with a stronger eyepiece. Challis agreed, but after dinner they delayed still further to drink cups of tea provided by Challis's wife, so that when they eventually made their way to the observatory the sky had clouded over.

Even this was not all. On 30 September the sky was clear, but there was strong moonlight and Challis felt disinclined to observe. John Russell Hind, director of George Bishop's private observatory in

Regent's Park, had heard of the discovery in a letter to him from Dr. Brünnow of Berlin; and using the 7-inch (17.78-centimeter) refractor in Regent's Park he not only located the planet, but resolved its disk.

Challis finally learned of the discovery on 1 October and started to analyze the 3150 star positions he had recorded since starting the hunt. He soon found that he had seen the planet twice in the first four days of his search, and, as he now expected, the object whose disk he had noted was also the planet. He wrote to Airy lamenting that "after

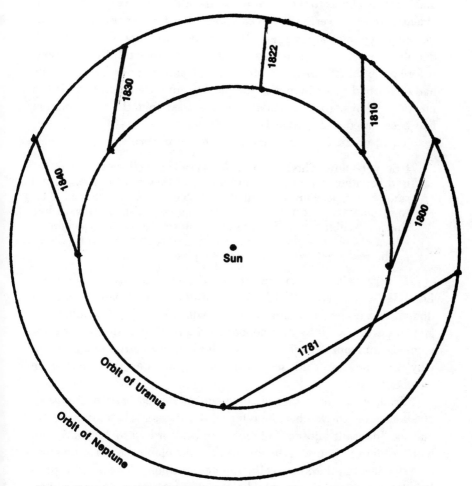

The pull of Uranus on Neptune. Before 1822, Uranus tended to pull Neptune back; after 1822, it tended to accelerate it.

four days of observing, the planet was in my grasp if only I had examined or mapped the observations."

Too late! The glory was Leverrier's, and both Airy and Challis became uncomfortably aware that they would be called to account. As yet the French and German astronomers had no idea that Adams had been involved at all. The first hint of the approaching storm came with a note by Sir John Herschel in the *Athenæum*, dated 3 October 1846: "A similar investigation had been independently entered into, and a conclusion as to the situation of the new planet very nearly coincident with M. Leverrier's arrived at (in entire ignorance of his conclusions) by a young Cambridge mathematician, Mr. Adams . . . who will doubtless, in his own good time and manner, place his calculations before the public." Challis wrote to Arago on 5 October saying that he had confirmed the existence of the planet, but once again he failed to mention Adams, so Arago naturally assumed that the confirmation had been based solely upon Leverrier's work.

Then, on 14 October, Airy wrote to Leverrier:

> I do not know whether you are aware that collateral researches had been going on in England and that they led to precisely the same results as yours. I think it probable that I shall be called on to give an account of these. If in this I give praise to others, I beg that you will not consider it as at all interfering with my acknowledgement of your claims. You are to be recognized beyond doubt as the real predictor of the planet's place.

If Airy's aim was to placate the French, he was disappointed. Leverrier was annoyed that his claim to priority had been even questioned, and he began asking questions which were highly embarrassing; in particular, why had nobody said anything until after the announcement from Berlin? Also, if Challis had been using Adams' data, why had he not said so when writing to Leverrier earlier in October?

François Arago was more vehement. At a meeting of the Academy of Sciences on 19 October, he referred to Adams's "clandestine" work and the "flagrant injustice" of any claim. According to Arago, Adams had "no right to figure in the discovery of the planet Leverrier, neither by a detailed citation, nor by the slightest allusion." The popular press in both France and England took the matter up, and feelings ran high. Airy did his best to pour oil on the troubled waters, but it was a long time before Leverrier and Adams were recognized as codiscoverers. Moreover, the Royal Society of London awarded its highest honor, the Copley Medal, in November 1846—to Leverrier.

Even the naming of the new planet led to controversy. Challis proposed "Oceanus"; others preferred "Janus." Leverrier suggested "Neptune," but later changed his mind and wanted the planet to be called after himself! It was months before "Neptune," favored by Galle, Encke, and most other leading astronomers became universally accepted.

The brightest point in the whole affair was when Adams and Leverrier finally met face to face in June 1847; they struck up an immediate friendship which lasted for the rest of their lives. They at least were free from bitterness—something which makes one suspect that Leverrier was not nearly so disagreeable as his colleagues often made out.

In retrospect, it is only right to say that both Airy and Challis were guilty of negligence. Airy was too narrow-minded to take much notice of the work of a young, unknown astronomer; Challis was incapable of adopting a procedure which would have brought any hope of quick success even when he had been spurred on to start the hunt. But there are other curious facts, also. The Rev. William Dawes knew about Adams's work; Lassell had been notified of it, even though the vital letter had been lost—so why was nothing done? And for that matter, why did not Adams make a search on his own account?

This last point has always puzzled me. I have not seen it mentioned elsewhere, and so, some time ago, I decided to check. I had not looked at Neptune for some months, and had only a rough idea of its position. Therefore, I took a star map, asked a colleague to mark in a position within a few degrees of Neptune's real place, and began to search, using my portable 3-inch (75-millimeter) refractor. I do not claim to be a keen-sighted observer, but it took me only ten days to identify Neptune by its movement alone. What I could do in 1976, surely Adams could have done in 1845. The mystery remains.

Moreover, it transpired that the discovery could have been made over half a century earlier. Lalande, one of the greatest of all observers, observed Neptune on 8 May 1795 and again two nights later. He noted a discrepancy—but merely rejected the first observation as being inaccurate. Why he did not check again is something which will never be clear. Had he done so, he could not have missed identifying the new planet.

Lassell, too, must have been disappointed. But for that sprained ankle, he would surely have taken action. As some slight compensation, he did at least discover Triton, the larger of the two satellites, on 14 October 1846, less than a month after Neptune itself had been found.

Neptune dealt a severe blow to the credulity attached to Bode's Law. The distance should have been 388 units; actually it turned out to be only 300.7, so that the law was completely shattered. Some astronomers still believe it to be significant, but on the whole it is thought much more likely to be due to coincidence.

Once again the Sun's family seemed to be complete, but lingering doubts remained, and throughout the rest of the nineteenth century there were occasional suggestions that another planet might exist, moving in the farthest regions of the Solar System. Eventually the problem was taken up by Percival Lowell, and the action shifted to the United States.

6

Early Investigations for a
Trans-Neptunian Planet

by Clyde W. Tombaugh

Most great discoveries in science are preceded by intuitions and followed by simple or crude methods, procedures, and use of inferior equipment. Often a succession of attempts take place in a progressive sequence, just barely missing the discovery. This was especially so in the case of the discovery of the ninth planet, Pluto.

Planet hunting is a slippery business. One small mistake can result in total failure; whereas in surveys made for cataloging stars, star clusters, and nebulae, a few mistakes scarcely detract from the value of the overall survey. Scientists are human. It requires a high level of alertness and thinking to avoid making blunders. In the case of Neptune, a very serious blunder was committed in the observational procedure by Challis.

At certain stages in scientific knowledge, new lines of research come forth almost simultaneously by investigators in different countries and unknown to each other until after the discovery is announced. Such was the case in the discovery of Neptune. In all of the sciences, one discovery leads to another, influencing the events that soon follow.

England lost the credit of picking up Neptune, which went to Galle of the Berlin Observatory a little later. As a result, Challis spent the remainder of his life in gnawing remorse. This tragic incident could have been prevented had Sir George Airy, the Astronomer Royal, taken John Adams's mathematical prediction seriously a year earlier while there was still time before Leverrier's analysis was completed. As a result, Airy reaped upon himself much severe criticism.

Some astronomers became so engrossed in a routine program of cataloging that they seem to have lost much of their alertness and intuition, or perhaps they had little of it to start with, and they were unable to cope with the unexpected. Sir George Airy seemed unable to realize the implications of Adams's calculations.

The most astounding case was that of the French astronomer, Lalande, in 1795. He was engaged in compiling a star catalog. On May 8 that year, he recorded the position of Neptune as a star. In one respect, he was a careful observer. It was his habit to recheck the positions of the stars in each field a few nights later to guard against errors. On May 10, he rechecked the field and found a discrepancy for one of the stars. It was Neptune that had shifted. He assumed that he had made an error in the measurement of the star's position two nights earlier and then dismissed the matter. Lalande could not have had a better clue or test for the identification of a new planet. Uranus had been discovered by Herschel only fourteen years earlier and it is remarkable that Lalande did not grasp the meaning of the displaced star image. Neptune was not again seen by the eye of man for another fifty-one years, until 1846 when Galle observed it in Berlin.

If Lalande had recognized the object as a moving new planet, all of the mathematical predictions of Adams and Leverrier would never have occurred in the 1840s. France would have received full credit for the discovery of Neptune. Equally strange, the planet Uranus was observed and recorded as a star at least twenty times before it was discovered by Herschel in 1781.

As early as 1834, Peter Andreas Hansen expressed the opinion, in correspondence with Alexis Bouvard, that a single planet would not account for the increasing "residuals" in the orbital longitude of Uranus. Residuals are the leftover deviations in position after the perturbation effects of the other known planets are taken into account. This thought was also the basis of Percival Lowell's theoretical investigations, some seventy years later. Lowell made careful studies of all the known planets and predicted a planet beyond Neptune. Hansen's idea of a planet beyond Uranus is a remarkable insight in that it preceded the actual discovery of Neptune by twelve years.

Three decades after the discovery of Neptune in 1846, the prospect of other distant planets began to engage the attention of several investigators. In 1877, David Peck Todd, from a graphical analysis of the perturbations of Uranus, predicted a planet at 52 astronomical units from the Sun. Todd assigned a diameter of 50,000 miles (80,000 kilometers) for his planet (considerably larger than Neptune). With this diameter and distance, the planet would have exhibited an apparent disk about the same size as does Neptune, equal to 2 arcseconds. Since astronomical photography was not effectively developed at this time, Todd chose to search for the planet visually with high magnifying powers of 400 and 600 diameters on the new 26-inch (66-centimeter) telescope at the U. S. Naval Observatory.

On thirty, clear, moonless nights, between 3 November 1877, and 5 March 1878, he searched a strip of sky along the Invariable Plane (the mean plane of the Solar System, inclined from the Ecliptic by only 1.6 degrees). The strip was 2 degrees wide and 40 degrees long. Todd examined each star for a perceptible disk. The examined strip lies in the constellations of Leo and Virgo far from the star-rich regions of the Milky Way. Nevertheless, it involved individual inspection of some three thousand stars down to the thirteenth magnitude—a considerable task. A similar strip in Gemini would have increased the work tenfold. He suspected many objects, which were reobserved on the following nights and again several weeks later. Since they did not move in position, they were dismissed. Todd had much confidence in the negative results of his search. Right he was, for there is no such planet.

Other investigators of possible planets beyond Neptune in the latter part of the nineteenth century were George Forbes of Edinburgh, Scotland; Camille Flammarion and Jean Baptiste Gaillot of France; and H. E. Lau of Copenhagen, Denmark. Several of the proposed planets were assigned considerable masses and at distances which would have rendered them brighter than Pluto. (These predicted planets would have been relatively easy to detect on plates taken later with the 13-inch (33-centimeter) Lowell telescope under the Blink-Microscope-Comparator examination at Flagstaff. I covered a very wide belt around the sky with great care [1929-43], and found no such planets.)

During the first three decades of the twentieth century, two investigators stand out: Percival Lowell (1855-1916), and William H. Pickering (1858-1938). Lowell's prediction of his Planet X was summarized in his *Memoir on a Trans-Neptunian Planet*, published in 1915. He predicted two possible planetary positions in almost opposite regions of the sky. Pickering published several papers about his hypothetical planets in *Popular Astronomy* magazine in the late 1920s. Pickering's Planet O was very similar to Lowell's Planet X, except that Pickering predicted a fainter magnitude of 15. As it later turned out, this value coincided with Pluto's actual photographic magnitude at the time of discovery.

Incidentally, Pickering assisted Lowell in setting up the Lowell Observatory in 1894 at Flagstaff, Arizona. Both were prolific observers of Mars. Pickering flitted from one idea to another, in contrast to Lowell's temperament of persistence and rigor. Pickering's drawings of Mars were rough and done with little care, which irritated Lowell. There was a real clash of personalities, and they parted company to later become bitter rivals.

Of special interest was Pickering's Planet P. In 1911, he assigned the planet a distance of 123 astronomical units with an orbital period of 1400 years. Then in 1928, he revised the characteristics of the proposed planet to a circular orbit at a distance of 67.7 astronomical units, with a period of 557 years and having 20 Earth masses. In 1931, he again revised his theory to predict a planet with an elliptical orbit at a distance of 75.5 astronomical units with a period of 656 years and having a mass 50 times that of the Earth. This is considerable when one remembers that the mass of Neptune is equal to 17.3 times that of the Earth. According to both the 1928 and 1931 characteristics, the magnitude of the planet would have been 11, bright enough to easily be discovered. He indicated its position to be about 13 or 14 degrees off the Ecliptic, north of the constellation Aquarius.

In 1931, that area fell within my search strip, so I kept a sharp lookout for Planet P. At magnitude 11, the body would have been over 6 magnitudes or 250 times brighter than the minimum detection possible with my plates and would have been conspicuous. I found nothing there. At this point, I became completely disillusioned with all of Pickering's planet predictions and all other planet predictions based on comet orbital aphelia (the farthest point from the Sun).

Pickering based the predictions of several of his planets on the orbital characteristics of certain comets. Although comets are quite sensitive to gravitational perturbations, comets are subject to thrust forces from gases escaping from their nucleus, as in a rocket jet. When comets are in the perihelion portion (nearest the Sun) of their long elliptical orbits, the warming by the Sun vaporizes the ices which break through a thin, weak crust in spurts. It is difficult to know how much of an orbital change is due to one or the other force. In the prediction of planets, only the gravitational force is involved.

In 1919, Pickering persuaded the director of the Mount Wilson Observatory to search the area predicted for his Planet O. The search was undertaken by Milton Humason with the observatory's 10-inch (25.4-centimeter) astrograph (photographic telescope). At that time, Pluto was right in the richest star region of the Milky Way in western Gemini. Four plates were taken. Pluto happened to lie just outside of the plate area subjected to the most thorough scrutiny, and the images of Pluto were not detected. This was an extreme disappointment to Pickering.

(I knew nothing about the 1919 Mount Wilson search until after my discovery of Pluto in 1930. After the announcement in March, in which the exact location of Pluto was given, Mount Wilson calculated backward in time the position of Pluto on their 1919 plates and found the Pluto images.)

In 1928, Pickering revised his Planet O orbit, such that its mean distance was about the same as for Neptune (30 astronomical units), but that its orbit crossed Neptune's orbit, about half of the circuit inside and half outside. (Actually, Pluto's orbit does cross that of Neptune, but only 19 years of the 248-year period lies within Neptune's orbit.) He assigned the mass to be 0.75 that of the Earth and a magnitude of 13.5.

The Harvard Observatory searched for Pickering's Planet O on three nights in January 1928, but without success.

In the March 1929 issue of *Popular Astronomy*, Pickering pleaded for someone to look for his planet. Again, there was observational apathy. By then the apathy was probably justified. Pickering had revised the characteristics of his several proposed planets so many times that there was little confidence left for would-be searchers.

In the same issue of *Popular Astronomy* was V. M. Slipher's annual report of the Lowell Observatory. This report indicated that a systematic search for a new planet was about to be resumed at Flagstaff with a new 13-inch (33-centimeter) telescope.

What caused this burst in predicting new planets? The mathematical triumph in the remarkable prediction of Neptune became a model. For one to be able to predict a new planet became a status symbol.

Many readers may prefer to skip over the next few pages which explain the fundamentals of orbits. But, anyone who has done reasonably well in high school geometry has a good chance of acquiring some understanding of it and appreciating the task involved in predicting a planet. The same fundamentals are involved in the navigational aspects of sending spacecraft to the various planets.

It is well to remember that prior to the age of electronic computers, complex computations were carried out by hand with the use of logarithm tables to the sixth decimal place. Now with the electronic computer, an orbit can be solved in a very small fraction of the time required with logarithms. But one must understand how to program the computer.

Kepler's three laws state quite simply the fundamental manner in which all bodies move in the Solar System and in multiple star systems, as well. It is perhaps pertinent to state them here, since all three are involved in predicting planets.

Kepler's First Law: Each planet moves about the Sun in an orbit that is an ellipse, with the Sun at one focus of the ellipse. This law applies also to the motion of asteroids and periodic comets.

Since an ellipse is a closed curve, it has two foci. A circle is a special case of an ellipse which has an eccentricity of zero, and the two foci coincide in the center.

Kepler's Second Law: The straight line joining a planet and the Sun sweeps out equal areas in space (that is, equal sector areas in the plane of its orbit) in equal intervals of time. The line joining the planet and Sun is variable in length. When the planet is farther from the Sun, the planet moves in its orbit more slowly, but since the joining line is longer, it still sweeps out the same area. Thus, it is very easy to calculate the speed of a planet at any point in its orbit.

Many comets are nonperiodic, that is, they never return to the Sun's neighborhood. Their orbits are open-ended and have only one focus (at the Sun). In the case of the conic section curve, which has an eccentricity of exactly one, it is a parabola. If the eccentricity is greater than one, it is a hyperbola. Kepler's Second Law of equal areas also defines the varying speed of such a comet in its trajectory.

Kepler's Third Law: The squares of the sidereal periods (of revolution) of the planets are in direct proportion to the cubes of the semimajor axes of their orbits. A sidereal period is the interval of time between successive rotations of a planet on its axis with respect to a given star, called a sidereal day; also of the revolution of a planet around the Sun with respect to a given star, called the sidereal year. For example, the Earth's sidereal day is 23 hours, 56 minutes, whereas the civil day (with respect to the Sun) is exactly 24 hours. The semimajor axis is the element that defines the size of the orbit. It is one of several elements solved in the computation of the orbit from three accurately observed positions, which are separated by adequate intervals of time over only a portion of its orbit. Thus, one can calculate the period of revolution, instead of having to wait until the planet completes one revolution.

In orbit computation, one solves for several unknown quantities. It is a very complicated mathematical exercise. One by one, the elements come forth in the later stages of the long computation. It is actually exciting to determine what the elements will be.

The eccentricity defines the shape of the orbit, whether a circle, an oval or elongated ellipse, or the open-ended parabola and hyperbola. Three elements define the orientation of the orbit. The inclination determines the tilt of the orbit plane in reference to the plane of the Earth's orbit. The longitude of the ascending node determines where the orbital plane of the body crosses the plane of the Earth's orbit from south to north. It is reckoned eastward from the vernal equinox point. The next element determines the longitude of perihelion (point in the body's orbit where it is nearest the Sun) measured in the orbital plane of the body, reckoned from the ascending node. Then one solves for the date that the body passes through its perihelion. With these fundamental orbital quantities determined, one can calculate backward in time the position of the body on a desired date

Explanation of the Orbital Elements for Encke's Comet

Encke's comet was chosen to portray the nature and meaning of the orbital elements of a celestial body in orbit around the Sun for two reasons. First, Encke's comet is the shortest known periodic comet (period = 3.3 years); therefore, the orbits of the Earth and the comet are more comparable in size to portray the orbital parameters. Also, while the orbits of the Earth, Venus, and Neptune are nearly circular, the higher eccentricity of Encke's comet better portrays the properties of the ellipse and is comparable to several asteroids and many comets.

Three planets have relatively high eccentricities: Pluto = .250, Mercury = .206, and Mars = .093; those of Jupiter, Saturn, and Uranus are about half that of Mars. Note that the eccentricity of the nearest planet to the Sun, Mercury, is almost as much as the most distant planet, Pluto.

The points "P" for perihelion (nearest to the Sun), and "A" for aphelion (farthest from the Sun), always lie along the line of the major axis of all orbits. The Sun is always at one of the two foci of an ellipse and always in the plane of the orbit for any tilt angle.

The circle just beyond "P" of Mars's orbit represents the mean distance of Mars from the Sun.

The shaded sectors in the orbit of Encke's comet are equal in area and illustrate Kepler's Second Law of equal areas swept by the radius vector. The arcs of the sectors show the orbital distance traveled by the comet in one month.

The semimajor axis, "a," defines the size of the orbit. The eccentricity, "e," defines the shape of the orbit and involves the ratio of the major and minor axes of the ellipse. The longitude of the ascending node, "Ω," is measured eastward from the vernal equinox, "γ," and defines where the comet crosses the plane of the Earth's orbit from south to north. The upper portion of the comet orbit hides that portion of the Earth's orbital plane. The longitude of perihelion, "ω," is measured from the "Ω." The inclination of the comet orbital plane to the plane of the Earth's orbit is denoted by "ι" = 12 degrees in this case. The latter three elements define the orientation of the orbit.

The orbit of the largest asteroid, "Ceres," is almost exactly at the average distance of the asteroids from the Sun.

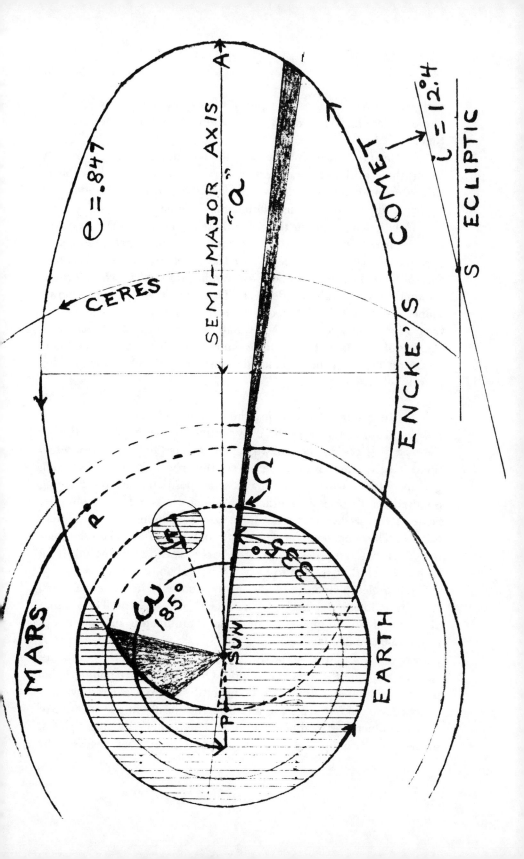

and year to recover the exact position on old photographs. One can also calculate forward exactly where the body will be at some future time.

To one familiar with the properties of orbital elements, by merely seeing these arithmetic numbers, a person can visualize the orbit in three-dimensional space with reference to the Earth's orbit and how it lies with respect to the constellations in the night sky.

In elliptical orbits in the Solar System, the Sun is at one of the foci, with nothing at the other focus. There are a great number of periodic comets which recede into the outer portions of the Solar System. It is known that there are families of comets whose orbital semimajor axes are almost identical. This suggests that some unknown planet lies near the other focus of their ellipse. Jupiter is known to have a family of some fifty comets. Saturn, Uranus, and Neptune have families of comets, but much fewer known members. As stated earlier, Pickering and others had predicted planets from comets with more distant aphelia.

In the history of predicting the existence of new planets, unless the prediction could be verified by actual observational detection in the vast sky, the prediction was open to question, especially when there were so many drastic revisions of the orbits. The task of singling out one little starlike image that shifted position from the teeming hundreds of thousands of other star images was a formidable one. Some of the predictors did not have access to the use of a wide-field camera. They depended on others with observing skills to do the finding. There was a reluctance by observers to spend time and effort on such a search that seemed to have a low probability of success.

On a clear, moonless night, away from city lights, about 3000 stars are visible to the unaided eye in the sky hemisphere. The number of stars visible with binoculars of 40 or 50 millimeters aperture jumps to 150,000 and 250,000 respectively. With a 6-inch (15.24-centimeter) telescope, one sees stars to the thirteenth magnitude and the number of stars visible increases to over 2 million. On blue-sensitive photographic plates, the number of stars recorded over the entire heavens down to a magnitude of 15.5 (the photographic magnitude of Pluto at the time of discovery) increases to about 20 million. On the Pluto discovery plates, the star density happened to be average, and the number of stars to the seventeenth magnitude was almost exactly 1000 stars per square degree! There were 162 square degrees on the 14-by-17-inch (35.56-by-43.18-centimeter) discovery plate! No wonder that predictors tried to place the planet within a few degrees of its actual position.

Neptune was detected visually, looking into a telescope, with scarcely more than a few hundred stars to be checked. But the great number of stars to be checked for a much fainter planet requires the use of photography.

7

Lowell's Investigations for Planet X

by Clyde W. Tombaugh

Percival Lowell (1855–1916) was the eldest of the illustrious family of Lowells in Boston. Percival's brother, A. Lawrence Lowell, served as president of Harvard University from 1909 to 1933. Lowell had three sisters. His youngest sister, Amy, was renowned as a cigar-smoking poet. Another sister was Elizabeth, the mother of Roger Lowell Putnam, who became the Lowell Observatory Trustee from 1927 until the early 1970s. Putnam played a key role in renewing the final search for Lowell's Planet X that led to Pluto's discovery.

Like all of the Lowells, Percival was well educated, traveled widely, had money, worked hard, and had the temperament of philosopher and poet. Percival Lowell started out in the diplomatic service and spent several years in Japan and Korea. There he gleaned a knowledge of the oriental mind and society, which he described in his book, *The Soul of the Far East.*

At an early age, he became interested in astronomy. The discovery of the line-like markings on Mars by Giovanni Schiaparelli in Italy in 1877, particularly captivated Lowell's interest. These linear mark-

Percival Lowell (1855–1916) mathematically predicted the existence of a trans-Neptunian planet and initiated the search that led to the discovery of Pluto in 1930 at the Lowell Observatory.

ings became known as *canals* through a mistranslation of Schiaparelli's reports. These unnatural features on the disk of Mars suggested to him that they were constructed by a highly civilized race of intelligent beings.

Lowell took a 6-inch (15.24-centimeter) telescope with him to the Far East to see for himself some of these strange things on Mars. When he heard of Schiaparelli's failing eyesight a decade later, Lowell resolved to take up a vigorous observational study of Mars and left his diplomatic career. First he conducted a survey for the best possible observing sites in several countries where the atmospheric conditions would permit the most effective observations for planetary studies, particularly for Mars.

Among his helpers were William H. Pickering and A. E. Douglass, both observers of Mars, and the latter noted also for his studies of past climates by tree-ring analysis.

Finally, Lowell decided to found his observatory at Flagstaff, Arizona, on the pine-covered plateau at 7250 feet (2210 meters) above sea level. The immediate spot was on a volcanic mesa one mile (1.61 kilometers) west and 350 feet (107 meters) above the then-small town of Flagstaff. He was able to borrow an excellent 18-inch (45.7-centimeter) refracting telescope for the favorable opposition of Mars in 1894.

In the meantime, he ordered a 24-inch (61-centimeter) refractor for his observatory. He specified a focal ratio in which the focal length was 16 times longer than the diameter of the objective lens. This ratio kept the telescope short enough to fit in the dome formerly built to house the borrowed 18-inch (45.7 centimeter). This unfortunate ratio reduced the resolving power needed to see finer detail.

With a simple convex lens, the red rays of light come to focus a little farther behind the lens than the blue rays do. An eyepiece is used to magnify the image formed by the larger objective lens. By moving the eyepiece inward, it can be made to focus on the image formed by the blue rays, whereas the images formed by the other wavelengths of light are out of focus and blurred. One may move the eyepiece a fraction of an inch farther from the objective lens and be focused on the yellow-green image, which is the best compromise because the retina of the human eye is most sensitive to the yellow-green. However, the eyepiece is now outside of the blue focus and inside of the red focus, and both the blue and red images are blurred and superimposed on the yellow-green image, which is in focus.

By adding a second lens, a concave component to the objective, but made of different kind of glass higher in dispersion power, the different wavelengths of light are reunited to white light without annulling all of the converging power of the convex member. This is called an achromatic objective. It makes possible excellent performance for lenses of fair size, provided that the focal ratio is long enough. Even so, an achromatic, which is a two-component objective lens, cannot bring the more extreme wavelengths of light to a perfect common focus. One must realize that a convex lens is actually a prism of varying apex angles, hence the dispersion of white light into the colors of the rainbow. By choosing a proper focal ratio for a given size achromatic lens, the effect of the secondary color or chromatic aberration can be brought to tolerable limits, where only a weak purplish-blue fringe is seen around the image of the object being viewed.

The larger the achromatic lens, the longer the focal ratio must be to meet the tolerable limit. The larger the lens, the more resolving power it will have for resolving fine planetary detail and close double stars.

During my years at the Lowell Observatory, I used the 24-inch (61-centimeter) refractor hundreds of times in viewing the Moon and planets. Also, I studied the optical characteristics of the telescope. The telescope was equipped with a large iris diaphragm just in front of the large lens, which could be controlled from the eyepiece at the lower end with a dial to indicate the size of the aperture being used. I experimented with various apertures, various power eyepieces, and color filters. The accuracy of the curves on the surfaces make this telescope one of the best in the world. At full 24-inch (61-centimeter) aperture, the focal ratio is 16, but the excessive color aberration is severe enough that one sees less than when diaphragmed to 20 or even 18 inches (50.8 or 45.7 centimeters). Even then the images of the Moon or planets tend to "wash out" if magnifying powers higher than 500 diameters are used.

Most of the time Lowell preferred to use the 24-inch (61-centimeter) refractor diaphragmed to a 16-inch (40.64-centimeter) aperture, and magnifying powers of 310 or 400, according to his observing records. This combination appears to improve the contrast of markings on a planet's disk but reduces the resolving power needed to see finer detail. When I used this combination I saw over a hundred of the so-called canals much as he drew them, as continuous lines. When I used a lower-power eyepiece yielding a power of 310, the canals appeared even finer. Why? The disk of the planet appears smaller but brighter with a lower power. The bright reddish-ocher desert areas on each side of a darkish stripe appear to bleed over onto the dark stripe by what is known as "irradiation." Thus the eye sees the dark stripe more narrow than it really is. This illusion was misleading to Lowell. Years later, when I was observing Mars in 1950 at the McDonald Observatory in west Texas, I persuaded Kuiper to use a 27-inch (68.6-centimeter) eccentric diaphragm on the 82-inch (208.3-centimeter) mirror. I had read the test report that the mirror's figure was not adequate to provide the theoretical resolving power of so large a mirror, assuming that one had super-seeing atmospheric conditions to permit it. He was using a magnifying power of 660 diameters. I said that the extremely bright disk of Mars with that low a power would obliterate all the finer detail from excessive irradiation. It did. I could see nothing of the stronger canals where I knew they were positioned. Finally, he agreed to try a cardboard diaphragm so positioned in the converging beam a few feet in front of the Cassegrain focus where we viewed the planet with an eyepiece. The circular hole in the cardboard rightly positioned gave the equivalent of a 27-inch (68.6-centimeter) circular aperture lying between the occulting shadow of the smaller convex Cassegrain mirror and the edge of the primary mirror and lying between the struts holding the secondary mirror.

The purpose of this was to dodge the adverse diffraction effects caused by the central obstruction of the Cassegrain mirror. The major portion of the 82-inch (208.3-centimeter) primary mirror was now blocked off, reducing the excessive brilliance of the disk of Mars to a tolerable level. The improvement was amazing. Not only could I see about twenty canals, but the canals now appeared as a series of individual, unconnected darkish spots lying in a general rough alignment. The seeing was extraordinarily good on that night and a few nights following. I would have preferred using a power of 800 if such an eyepiece had been available. Since this was an all-mirror telescope, except for the eyepiece, there was a complete absence of color aberration.

It would have been much better if Lowell had specified a focal ratio of about 25 for his 24-inch (61-centimeter) refractor and built a new and larger dome. He would then have been able to use his refractor to full resolving capacity. Since Lowell saw the canals as straight, fine lines, suggesting artificiality, he developed his famous ideas about Mars. Of course it conflicted with what other astronomers, who used somewhat larger telescopes, saw. Had he perceived the real nature of the canals, he would have dispelled the notion of the artificial appearance of the canals, and the whole episode of the Mars canal controversy might have been spared. Thus, Lowell missed discernment of the real nature of the canals by only a factor of about 1.5!

There was also another factor. Lowell had an ideological axe to grind. He hated war, as can be seen from the last chapter of his book, *Mars and its Canals*, published in 1906 by the Macmillan Company.

That Mars is inhabited by beings of some sort or other we may consider as certain as it is uncertain what those beings may be. Girdling their globe and stretching from pole to pole, the Martian canal system not only embraces their whole world, but is an organized entity. Each canal joins another, which in turn connects with a third, and so on over the entire surface of the planet. This continuity of construction posits a community of interest. Now, when we consider that though not so large as the Earth the world of Mars is one of 4200 miles diameter and therefore containing something like 212,000,000 of square miles, the unity of the process acquires considerable significance. The supposed vast enterprises of the Earth look small beside it. None of them but become local in comparison, gigantic as they seem to us to be.

The first thing that is forced on us in conclusion is the necessarily intelligent and non-bellicose character of the community which could thus act as a unit throughout its globe. War is a survival among us from savage times and affects now chiefly the boyish and unthinking element of the nation. The wisest realize that there are better ways for practicing heroism and other and more certain ends of insuring the survival of the fittest. It is something people outgrow. But whether they consciously practice peace or not, nature in its evolution eventually practices it for them, and after enough of the inhabitants of a globe have killed each other off, the remainder must find it more advantageous to work together for the common good. Whether increasing common sense or increasing necessity was the spur that drove the Martians to this eminently sagacious state we cannot say, but it is certain that reached it they have, and equally certain that if they had not they must all die. When a planet has attained to the age of advancing decrepitude, and the remnant of its water supply resides simply in its polar caps, these can only be effectively tapped for the benefit of the inhabitants when arctic and equatorial peoples are at one. Difference of policy on the question of the all-important water supply means nothing short of death. Isolated communities cannot there be sufficient unto themselves; they must combine to solidarity or perish.

From the fact, therefore, that the reticulated canal system is an elaborate entity embracing the whole planet from one pole to the other, we have not only proof of the world-wide sagacity of its builders, but a very suggestive side-light, to the fact that only a universal necessity such as water could well be its underlying cause.

For several months around opposition time, Lowell was at Flagstaff observing Mars. At least half of the time, Lowell lived in Boston. He went on lecture tours, proclaiming his theories and proof of intelligent life on Mars and deriding the shortcomings of Earth men in their international global squabblings. Lowell wrote several popular books on Mars describing the behavior of the seasonal phenomena and his deductions as to their meaning.

Even though I did not share Lowell's interpretations, I found his books delightful to read. He wrote in a style of poetic prose. Much of his description on the appearance of various features and Martian seasonal changes were essentially accurate and correct.

Lowell's theories regarding the so-called canals soon came under severe criticism by other astronomers in the United States and Europe. Most of them had little experience in planetary observing. However, several French astronomers were experienced Mars observers and had more powerful telescopes. They could not agree with Lowell's description of the canals. Lowell replied to his critics in counter-criticisms. The controversy grew worse and worse, and the Lowell Observatory became virtually an outcast in professional astronomical circles.

The prestige of mathematically predicting a new planet apparently was sought by Lowell in order to gain more respectability for his theories about Mars. However, an interesting coincidence steered Lowell into an investigation for a possible trans-Neptunian planet.

As observations of Neptune began to accumulate after its discovery, Adams's and Leverrier's mathematical predictions were challenged as "a happy accident." Although the predicted orbits for Neptune by Adams and Leverrier were quite similar, both orbits were much be-

yond Neptune's actual orbit. Apparently, both Adams and Leverrier
were unduly influenced by Bode's Law. Indeed, if Neptune could be
eliminated, Pluto's mean distance from the Sun would fit into Bode's
Law very well. But it is at Neptune's orbit where Bode's Law first
breaks down. That Neptune was found so close to the predicted positions of both Adams and Leverrier is regarded by some as extraordinarily good luck. The Neptune controversy is described in Grosser's
excellent book, *The Discovery of Neptune* (1962).

It is of interest that Lowell based his prediction of a trans-Neptunian planet at first on extrapolations from the observed association
of the orbits of comets and meteor streams with the orbits of the outer
planets, as others had done. After abandoning the comet method for
predicting a planet, Lowell needed data on the residuals of Uranus
and Neptune. 'Residuals' are the leftover deviations in position after
the perturbation effects of all of the other known planets have been
taken into account.

Lowell's search for what he came to call "Planet X" began early in
1905. A special 5-inch (12.7-centimeter) aperture photographic objective was selected and procured from the John A. Brashear Company in
Pittsburgh. This was mounted equatorially for tracking over a period
of 3 hours on long exposure times to record the faintest stars possible.

This series of 440 photographic plates covered a relatively narrow
zone around the sky, centered along the Invariable Plane (the mean
orbit plane of the planets). It is inclined to the better-known Ecliptic
by only 1.6 degrees. Although the plates covered a larger field width,
the Brashear camera gave sharp star images over an area of only 5
degrees. Beyond this central portion, the magnitude reach fell off
toward the edge. Accordingly, the plate regions had to be spaced every
5 degrees along the Invariable Plane. In the central portion, the plates
recorded stars to the sixteenth magnitude, which is 10 magnitudes, or
10,000 times fainter than the unaided eye sees on good, moonless
nights away from city lights.

Lowell examined these pairs of plates by laying one plate over the
other with a slight offset and scanned myriads of star images with a
hand-held magnifier. The stars are so far away that they appear rigidly
stationary and "fixed" during an interval of several years on a small-
scale pair of plates. Lowell was looking for one that would show a
small shift, compared to the more rapid shifts of asteroids. This means

of examination was unsatisfactory and uncertain. At that time, there was no Blink-Microscope-Comparator, which I used in my discovery of Pluto.

As we know now, Pluto was in a more remote part of its orbit during the 1905–1907 search. Pluto's photographic magnitude then was 16.0, which was a half magnitude fainter, or about two-thirds as bright as it was at the time of its discovery in 1930. Also, Pluto was just outside of the Invariable Plane strip because of its orbital inclination. Consequently, Lowell did not have a chance of success in the 1905–1907 photographic search.

At the opposition point (180 degrees from the Sun), all Solar System bodies, including comets, beyond the Earth's orbit are shifted westward (apparent retrograde). In that position they are getting the full tangential effect of the Earth's more rapid orbital parallactic motion. It is the same effect as when one passes a car on the highway, and the other car appears to drift backward with respect you, even though it is traveling in the same direction as you are. Approximately, the amount of apparent retrograde shift is inversely proportional to the object's distance. Thus, one has a continual running parallax created by the Earth's orbital velocity, by which one can estimate almost instantly the distance of the planet suspect.

The angular distances of the apparent stationary points from opposition for Mars and Jupiter are, respectively, 36 and 64 degrees. Since most of the thousands of asteroids lie between the orbits of Mars and Jupiter, these sectors on either side of the opposition point must be rigorously avoided. Any asteroid near its stationary point will exhibit the small shift characteristic of a trans-Neptunian planet. The apparent stationary points of very distant planets lie at greater angular distances from the opposition point. The stationary situation is created when the tangential vectors of the Earth's and an exterior planet's motion cancel out. Contrary to general supposition, the looked-for shift to detect a distant planet is not due to the planet's actual orbital motion, but the Earth's daily parallactic motion instead. This is evident when one remembers that all planets move eastward in their orbits. What we look for is the apparent retrograde shift to the westward. Thus, each region of the Zodiac must be photographed at the proper season. For instance, regions in Gemini must be photographed in January, not in March, nor in November. The more powerful the search telescope, the more hundreds of asteroids are recorded and the greater the number of deceptive planet suspects.

A little arithmetic will show why the observing program was so arduous. The plates should be taken within 20 degrees of the "opposition point" to avoid confusion with the numerous asteroids. But the opposition point moves eastward 30 degrees each month because of the Earth's revolution around the Sun. Since the long-exposure plates cannot be taken under conditions of moonlight, observing hours are limited to about 2½ weeks of each month. The opposition point is on the meridian at midnight. In order to avoid magnitude loss from atmospheric absorption and differential refraction, the plates must be taken within a few hours on either side of the meridian. With 5-degree spacing of the plates, at least six plate regions had to be covered each month. There must be a pair of plates for each, or a total of twelve plates. Each plate had to be exposed three hours—a total of thirty-six hours of actual exposing.

It is best to have three plates of each region for checking planet suspects. Now the total is fifty-four hours of net exposing. This does not include the time for resetting to another camera region, nor for changing plate-holders.

More things can go wrong during a three-hour exposure than a one-hour exposure. According to Murphy's Law: Anything that can go wrong, will go wrong. Unexpected clouds may come in after two hours of painstaking guiding, or the clock drive freezes, or the observer inadvertently bumps the camera (because the Brashear camera was not on a heavy, sturdy mounting). Still another hazard is always present. The observer may obtain good first plates, then a spell of prolonged bad weather prevents getting the duplicate plates before the Moon comes around into that part of the sky. In all of these situations, the observer must do the work over again. So it is necessary to try again next month. By that time, some of the plate regions are too far from the opposition point. The observer tries to retrieve them anyway. The plate scanner runs into an asteroid near the stationary point. "Ah! The planet," great excitement! Then comes an urgent telegram from Boston, "Rephotograph the region." "Can't, moonlight in the way." Or, if the Moon has moved sufficiently to the eastward out of the way, a spell of cloudy weather has ensued. By this time, the photographic work of the next lunation (the long-exposure photographic portion of the month) is at hand. When the skies clear, the observer must interrupt the work-run in the new opposition region. Likely, the project is already behind schedule due to some kind of bad luck. The observer rephotographs the old region in duplicate. Plates are sent by train mail from Flagstaff to Boston. Soon the verdict: "Object is gone—obviously, it was an asteroid," which had sped away out of the immediate region. This kind of wild-goose chase could have been avoided if all of the plates had been taken closer to the opposition point and in triplicate. It was a hectic ordeal.

A two-hour exposure of vigilant guiding taxes the observer considerably, but a three-hour one wears an observer out. In wintertime especially, the observer can become dangerously numb from the cold in an unheated dome in three hours without a break.

One gains only about one-third of a magnitude in prolonging the exposure from two hours to three hours. It would have been better strategy to have cut the exposures to two hours, which would have permitted time to take plates in triplicate. Also, this would have reduced the number of plate casualties, particularly from interruptions by clouds. Considerably more matched pairs of plates could have been obtained during the photographic run in each month, and falling behind the opposition point would have been avoided.

In 1909, Lowell acquired a 40-inch (101.2-centimeter) aperture reflecting telescope for his observatory. He had hoped to gain greater observational advantage for his studies of Mars. But the results were disappointing because they had set up the telescope in a subcellar pit, with the floor about 5 feet (1.5 meters) below ground level. The revolving dome cleared the ground by scarcely 1 foot (.3 meters). The dome was constructed of wood rib-work, covered by chicken wire, and then canvas. It was always a task to clear off the snow.

In the 1939 opposition of Mars, I assisted Dr. C. O. Lampland in taking hundreds of plates of Mars at the 80-foot (24.4 meter) Cassegrainian focus. The purpose of this program was to obtain photographs of Mars in different wavelengths of light over a much greater range than could be done with the 24-inch (61-centimeter) refractor.

I could see that the optics were too near ground level, which produced turbulence in the beam, resulting in less sharp images. I asked Lampland why they mounted the telescope in that manner. His response was somewhat evasive and I got the impression that I was treading on somebody's toes. Apparently, it was one of the observatory's sorrows, so I pursued the subject no further. Perhaps someone in the earlier days thought this plan would keep the temperature of the mirror more constant. Because of the less satisfactory performance of the 40-inch (101.2-centimeter) reflector, the older men were forever prejudiced against the use of reflecting telescopes for large-scale planetary observations.

Lowell got his 40-inch (101.2-centimeter) reflector because some of his critics challenged his Mars observations on grounds that his 24-inch (61-centimeter) refractor lacked sufficient resolving power to discern the real nature of the canals. But he soon lost interest in it, following the early unsatisfactory performance. Lampland once told me that Lowell seriously contemplated building a 50-inch (127-centimeter) aperture refractor for his Mars studies! This would have required a dome at least 100 feet (30.5 meters) in diameter. This ambitious dream must have dissipated into thin air quickly, after he figured out the enormous cost.

Lampland was the one who was stuck with the 40-inch (101.2-centimeter) reflector. Its principal use was the photography of star clusters, nebulae, and galaxies at the Newtonian focus. Beginning in 1922, he used the reflector at times in radiometric observations for measuring temperatures on the planets.

Lowell depended heavily on Lampland for the planet searching done in the first half of the second decade. From the many conversations I had with Lampland in later years, it was evident that Lampland greatly admired Lowell, in spite of Lowell's impatience at times.

After much computational work, using the residuals of Uranus, Lowell felt that he had indications where to search for another round —in the constellation of Libra. I am sure that the old 5-inch (12.7-centimeter) Brashear camera had long since been dismantled. Also, Lowell wanted to search to a fainter magnitude.

The only available instrument that could be used for this task was the 40-inch (101.2-centimeter) reflector at the Newtonian focus. In the early 1930s, both Lampland and E. C. Slipher told me much about the problems they had in an observational search program with this instrument.

The planet search with the 40-inch (101.2-centimeter) reflector began in 1911. However, the examination of these plates was greatly facilitated by the acquisition of a Blink-Microscope-Comparator from the Carl Zeiss Works in Germany. It was this Blink-Comparator that I used heavily from 1929 to 1943 in my extensive search and with which I discovered the images of Pluto.

The 40-inch (101.2-centimeter) reflector had a focal ratio of 5½. As in all reflecting telescopes of this type, the tangential coma limits the useful field of images to a 4-by-5-inch (10.2-by-12.7-centimeter) plate. With the focal length of the 40-inch (101.2-centimeter) telescope, the scale was 3.8 inches (9.65 centimeters) per degree. So each plate covered only 1 square degree of the sky (only 5 times the apparent area of the Moon). Such a search required taking a great number of plates. However, with the great light-gathering power of the 40-inch (101.2-centimeter) telescope, stars of the seventeenth magnitude could be recorded in less than ten minutes of exposure. Thus, much time was spent in resetting the telescope to another plate region and finding a suitable guide star.

After finding nothing in a specified area, Lowell and his chief computer, Miss Williams, would reexamine their computations and revise somewhat a new area to search.

After about a year, they gave up the search with the reflector. Lowell was able to borrow a 9-inch (22.9-centimeter) aperture, wide-angle camera from the Sproul Observatory at Swarthmore College. It was Dr. John A. Miller, director of the Sproul Observatory who loaned the doublet camera. It was this same Dr. Miller who came to Flagstaff in April 1930 to guide the computation of the Lowell orbit.

Dr. Lampland supervised the new search. The plates were taken by T. B. Gill and E. A. Edwards. From April 1914 to July 1916, nearly one thousand plates were taken over a considerable sky area. The plates were duplicated at two-week intervals. For a camera of these dimensions, it would have been better to use a one-week interval at opposition for more efficient blink examination.

In 1915, Lowell drastically revised the preferred region of search to longitudes of around 85 degrees in the extreme eastern portion of the constellation Taurus. This region is right in the heart of the Milky Way, and the plates taken were extremely rich with star images. Pluto was in this region then. In the examination of these plates, the Pluto images were missed.

After the discovery of Pluto in 1930, the better orbit computations

C. O. Lampland shared with Percival Lowell the 1914–16 trans-Neptunian search. Here he holds a device for measuring planetary surface temperatures. (*Photo by Jerry McLain*)

gave close approximations of where to retrieve the Pluto images on old plates. Lampland found the images of Pluto on one-hour exposure plates taken by Gill on 19 March and 7 April 1915. The images were quite weak, which was to be expected.

Because of Pluto's elliptical orbit, the planet had moved in appreciably closer than its distance during the 1905–1907 search. Also, Pluto was approaching its ascending node and was not as far south of the Ecliptic.

However, one aspect of this search shocks me. The dates of these plates show that the region was photographed 90 degrees from opposition, just barely past Pluto's apparent stationary point. This deprives the observer of ascertaining the distance from the Earth's daily parallactic motion. This kind of procedure is wrong in planet searching.

Lowell's memoir on a trans-Neptunian planet had disclosed two possible solutions, about 180 degrees apart and two sets of elements; particularly, semimajor axis of 43.0 and 44.7 astronomical units, and eccentricity of .202 and .195. The heliocentric longitude as of July 1914 was equal to 84 degrees and 262.8 degrees.

When the 1914 value of 84 degrees is brought up to 1930, the longitude of position was only some 6 degrees from where Pluto was found. This is what is so amazing. Also, Lowell indicated a mass between Neptune's and the Earth's; and a visibility of 12–13 magnitude according to albedo; also a disk of more than 1 arc second in diameter. Compared to Pluto, these latter values were badly off. He estimated the inclination of orbit at about 10 degrees.

Lowell had devoted considerable time and effort over a period of 13 years to the mathematical theory and search for a trans-Neptunian planet. Little was known to the outside world of his laborious work until after the Pluto discovery announcement in March 1930.

The failure to find Planet X was very disappointing to Lowell. In addition, World War I was a great source of pain to Lowell, a pacifist. On 16 November 1916, Lowell died of a stroke at the age of sixty-one, a discouraged and exhausted man.

This brought the planet search to a complete halt for thirteen years.

8

The Search Continues

by Clyde W. Tombaugh

World War I, the frustration of not finding Planet X, and Lowell's sudden death shattered the Lowell Observatory's efforts to do more planet search work. Lowell had done all that mathematical theory could do. The remaining task was clearly an observational one. They would have to wait until a more powerful wide-angle camera could be acquired.

Lowell left an endowment for his observatory so that the intensive studies of the planets might continue. He left behind three devoted assistants who would carry on this work. V. M. Slipher became director. He was a real pioneer in spectrographic studies, examining the chemical content of planetary atmospheres and radial velocities of globular star clusters. He was, also, the first to measure the high recessional velocities of galaxies. C. O. Lampland became assistant director, supervisor of the Planet X searches, photographic observations of variable stars, variable nebulae, novae, and radiometric observations for determining the temperatures of the planets. E. C. Slipher conducted visual and photographic observations of the planets and became a world-famous authority in the study of Mars.

V. M. Slipher (left), director of the Lowell Observatory during the final search for Planet X. To his right is his brother, E. C. Slipher, famed for observing Mars and other planets. (*Photo by Clyde Fisher*)

Despite Lowell's bequest, Lowell's widow tried to break the will regarding the estate. This bitter litigation dragged on for over ten years, robbing the resources of the estate through court costs and excessive attorney's fees. This resulted in a severe curtailment of funds for publications and especially prevented the acquisition of a more powerful and suitable instrument to renew any search for Planet X.

Guy Lowell, who had become the observatory trustee, did purchase 13-inch (33-centimeter) unfinished disks of glass from the Reverend Joel Metcalf's widow in 1925. But the long, drawn-out suit by Mrs. Percival Lowell had so impoverished the observatory's funds that there was no money to pursue the completion of that telescope. The staff felt it was futile to attempt any more search work until a more powerful instrument could be acquired.

Guy Lowell died in 1927, and Roger Lowell Putnam, Lowell's nephew, was appointed trustee. Roger Putnam had studied mathematics at Harvard and was well prepared for his new responsibility. Also, he was a very successful businessman and established a large package machinery plant in Springfield, Massachusetts. Naturally, he was eager to have the search resumed in hopes that his "Uncle Percy's" planet might be found.

Uppermost in Putnam's mind was the completion of a 13-inch (33-centimeter) objective lens, the design and construction of an equatorial mounting, and the building of a revolving dome to house the instrument. By building the mounting in the observatory shop and the dome on the observatory grounds to save money, they hoped to do the job for $10,000. This sum seems pitifully inadequate in terms of today's values and would approach $100,000 with today's (1980) costs.

Roger Putnam commenced a search for funds. Finally, he persuaded his uncle, Dr. A. Lawrence Lowell, then president of Harvard University to underwrite the project of completing the new telescope for the sum of $10,000. Story has it that Dr. A. Lawrence Lowell never visited his elder brother's observatory in Arizona, and Roger Putnam probably was not sure his uncle would be interested.

It was Putnam's enthusiasm that contributed so much to the resumption of the planet search. Putnam was to serve exceedingly well as trustee for four decades.

With funds assured, V. M. Slipher plunged into the work on the design for the mounting. From their previous experience with search equipment, Slipher incorporated features that would result in high-quality performance. One was a heavy, stable mount that would withstand gusts of wind during the long exposures. A large worm wheel, 4 feet in diameter, permitted easy and precise guiding. A long polar axis, supported at the upper and lower ends on separate piers, permitted exposures to be continued past the meridian.

This superb mounting was constructed by Stanley Sykes and his son, Guy. The dome was constructed by Mr. Mills, a Flagstaff carpenter who did a beautiful job.

The very important task was to find a highly competent optician to finish the objective lens.

Immediately after his return to Springfield, Putnam began looking for an optician who could do the precise figuring on the triplet lens (six optical surfaces) within a cost estimate that they could afford. Putnam contacted at least three optical telescope firms.

One was C. A. R. Lundin, of the old Alvan Clark firm in Cambridgeport, Massachusetts. Lundin gave an estimate and a contract was made with him.

Later, Lundin ran into an unexpected problem. One of the lens components had been ground so thin by Metcalf that Lundin had great difficulty during the final polishing in getting the precise figure, or curve, delaying completion.

On 22 January 1929, Putnam informed Slipher that Lundin had the lens components finished and mounted in a duralumin cell, ready to ship to Flagstaff. Then came the shocking bill of nearly $6,000. Lundin explained the reason for the increased cost.

Putnam promptly sent Lundin a check for the final payment without consulting Slipher. This upset Slipher because he felt that the lens should have been tested before final settlement was made. However, Slipher's fears were soon to be relieved after the first test plates

The 13-inch (33-centimeter) Lawrence Lowell telescope, with which my final extensive planet search was made. (*Courtesy Lowell Observatory*)

were made. Indeed, the quality of the star images and breadth of field exceeded their highest expectations.

The 13-inch (33-centimeter) objective lens arrived in Flagstaff on 11 February 1929.

Then began the final assault on the unknown in the outer regions of the Solar System. This is where I came into the story, for I had been brought to Flagstaff to work with the 13-inch (33-centimeter) telescope. In fact, it was not until I had arrived at the observatory that January that I learned what the photographic work mentioned in a letter from Dr. V. M. Slipher would be; namely, a renewed search for a trans-Neptunian planet.

It was V. M. Slipher himself who met me at the train station upon my arrival in Flagstaff, Arizona, in January 1929. Slipher took me up Mars Hill on a steep road with plenty of hairpin turns, all covered with wet snow. At a steep rise, he lost traction and had to back down the hill for another run at it. I must confess that I was a little alarmed.

He stopped in front of the administration building, and we went in and met Dr. Lampland and the secretary, Mrs. Fox. We chatted a little while. Then he took me upstairs. "This is your room. You may want to rest a little." I know I must have looked weary.

"If you care to come down a little before five, you can ride downtown with Mrs. Fox and find a cafe for dinner," he said, and left. I sat down in the chair almost in a state of shock. I was in another world, so very strange to me. I could feel the altitude of 7000 feet. I felt rather unnerved by it all.

After eating my dinner, or half of it, which I thought must have been bear meat, I walked the 1½ miles (2.4 kilometers) up the 350-foot (106.7-meter) climb to the observatory. The high, snow-covered peaks and the pine forest were truly beautiful. Dr. Lampland was in the secretary's office pounding out some letters on the typewriter. He was always trying to catch up on observatory correspondence that the secretary was unable to finish during the day. I was to find that this was habitual on evenings he was not observing, and this would go on for years. Dr. V. M. Slipher was in his office working on something. He was always there until midnight or after, except when he was running long-exposure spectrograms with the 24-inch (61-centimeter) refractor. In the reading room were marvelous books and jour-

Gateway to the Lowell Observatory on the old road, as it was in January 1929.

nals and the tick-tock sound, in synchronism and out, as the sidereal clock gained a second of time over the civil time clock. I was to hear this rather pleasant beat for the next fourteen years.

I was beginning to get sleepy. The altitude does this to you for a few months until you get acclimated. So I said good night. Dr. Slipher said, "Mr. Jennings [the handy man] comes here to the building about a quarter to eight to take the mailbag down to the post office and do some errands. You could meet him then, ride down to get your breakfast, and ride back with him at nine o'clock." This was to be my habit, when I did not work in the dome too late, for the next two years. I had a lot of adapting to do.

Accordingly, the next morning I rode down Mars Hill with Mr. Jennings in the observatory's old Model T Ford, ate my breakfast, and returned with him. A little later, Dr. Slipher had Mr. Jennings take me out to the new 13-inch (33-centimeter) telescope dome to see the new telescope. The telescope tube was open at the front end, where the 13-inch (33-centimeter) objective lens would be attached when it arrived from the East a few weeks later. We found Stanley Sykes and his son, Guy, drilling holes and tapping them in parts of the equatorial mounting for some of the hand controls. The polar axis consisted of a long, steel beam of welded right-angle and strap steel, and was unpainted. Several weeks later when the telescope was completed, it would be my lot to smooth up the welding rough spots and give the telescope a coat of red paint.

The observatory had visitor hours from half-past one to two o'clock every weekday, in which they would exhibit and explain astronomical photographs in the showcases in the library room, then take the visitors up an inclined sidewalk about 60 to 70 yards (55 to 64 meters) to show them the 24-inch (61-centimeter) refractor in the dome. It had been Mrs. Fox's duty to conduct these tours on a daily basis. She invited me to go along. The photograph transparencies of the Moon, planets, nebulae, and galaxies were beautiful. Then when she opened the door to the dome, I was awestruck at that huge telescope, 32 feet (9.8 meters) long. It was a magnificent sight to me.

A few days later, I was assigned the daily visitor tours, which I was to continue for several years. Indeed, over the years, I was to show several thousand visitors views through the 24-inch (61-centimeter) refractor.

In a few days, I was assigned an additional duty. Dr. Slipher took me down to the furnace room to show me how to throw several large, split pine logs into the furnace, which heated the large administration building. Each log was about the size of a railroad tie. About every two hours, somebody would go down to the furnace room to throw on half a dozen big, pine logs. Now Dr. Slipher assigned this job to me. About midnight, with the last log stoking, about 150 pounds of coal were shoveled in between the logs and furnace door and banked for the remainder of the night. Jennings was the first to arrive in the morning. It was his job to restoke the furnace. There was a sufficient amount of hot coals to ignite the new logs. Banking the furnace at night was my job for the next several years. Since I was the only one residing in the building, it was convenient for me to run downstairs and attend to it.

Later when I was making the photographic night runs with the

13-inch (33-centimeter) and slept until noon, the others would stoke the furnace. No one was above getting his hands dirty.

Over the years, I must have thrown several hundred tons of logs into that furnace. The building was a large one to heat, and the winters were about seven months long at this altitude of 7250 feet (2210 meters) above sea level.

This building housed ten office rooms, a reading room, the large library room (all on the first floor), and three darkrooms, the chemical room where plate developers were concocted, the plate-drying room, and two instrument rooms, all in the basement. The furnace room, log room, and coal bin were in a subbasement in a wing.

On the second floor, there was a conference room, a projection room, and several spare bedrooms. In the attic, which ran the whole length of the building, experimental apparatus, equipment, and junk parts covered most of the floor—an accumulation of over two decades. The attic and about half of the second floor were blocked off to save heat.

The other two telescope domes and the machine shop were all about 200 yards (183 meters) from the administration building.

Flagstaff is notorious for heavy snowfall. The total winter snowfall ranged from 6 to 15 feet (1.8 to 4.6 meters). It was common for a single snowstorm to dump 1 to 2 feet (.3 to .6 meters) and occasionally 3 feet (.9 meters) or more. Lampland's dome housing the 42-inch (106.7-centimeter) reflector had wooden ribwork covered by chicken wire and canvas. After every snowstorm, the snow had to be pushed off with a long wooden hoe. Lampland would never trust a laborer to get on top of his dome to do this, because one had to step where the supporting beams were; otherwise, one's feet would tear the old canvas. Lampland would do this himself, but he was now in his midfifties and perhaps not so surefooted as he once was. So this became one of my jobs for many years.

In the meantime, I was instructed in darkroom procedures, in preparation for my photographic work with the 13-inch (33-centimeter) telescope.

It was several weeks later that I met E. C. Slipher, who had been serving as a state senator in the Arizona legislature in Phoenix. Previously, he had served as mayor of Flagstaff. "E. C." was the Mars man.

I had many good looks at Mars with the 24-inch (61-centimeter) refractor. Many times, E. C. Slipher would come back to the administration building with a bunch of exposed plates and tell me, "Mr. Tombaugh, I took the camera off and put on the eyepiece, go on up and look at Mars while I develop these plates and reload the plate-holders." I was in seventh heaven. I made several sketches. E. C. Slipher would frequently make visual sketches to supplement the photographs. Oh, how that man could draw! I never saw any published sketches that beat his. Frequently, he engaged me in long chats about the planetary work. My interest pleased him greatly. This was to continue as a long-time kinship.

Dr. C. O. Lampland's office was just across the hall from mine. Frequently, he told me various anecdotes about Lowell, so that I almost felt that I knew Dr. Percival Lowell personally. Lampland said that Lowell urged them to engage in nonplanetary research also and publish papers in order to gain some credibility among stellar astronomers. Lampland pursued studies of variable stars, novae, and nebulae with the 42-inch (106.7-centimeter) reflecting telescope. V. M. Slipher investigated various objects with the spectrograph, particularly radial velocities of globular star clusters and nebulae. Indeed, V. M. Slipher was a pioneer in finding the very high radial velocities of galaxies, which led to the theory of the expanding universe.

But the long legacy of the Mars canals controversy had seriously demoralized the staff. A few times the two Sliphers and Lampland poured out to me the anguish in their souls. One time, Dr. V. M. Slipher told me, "The lot of the older men here is not a happy one," because of the ostracism from the astronomical community. It rather baffled me as to why they would tell these things to a newly joined underling. For a few years I was the only junior astronomer there.

Dr. V. M. Slipher, director of the Lowell Observatory, was having his dilemma also. He was responsible for having acquired an expensive new telescope and was feeling the pressure of having to produce significant results. The staff of three astronomers were too busy with other programs of research. That was why I was hired—to operate the new telescope and its demanding program. The long litigation had left the observatory seriously limited in funds. In those days, there were essentially no government grants, and private grants were few and small.

9

The 13-Inch Telescope

by Clyde W. Tombaugh

On 11 February 1929, the 13-inch (33-centimeter) objective lens arrived in Flagstaff. Jennings went down to the Santa Fe depot to load the large box and bring it up the 350-foot (106.7-meter) hill to the observatory, using the observatory's worn-out Ford Model T truck. I am sure V. M. Slipher must have accompanied him. Lampland and I were notified of its arrival at the administration building. We accompanied the entourage to the dome for opening the box.

The mood was tense. A screwdriver was never more cautiously used. With the top cover removed, Slipher cautiously reached down through the packing. "Well, it seems to be all in one piece," he said.

The packing was removed, and the 13-inch (33-centimeter) objective in its cell was placed on the table. There was our beautiful jewel. Then Slipher spied the focusing slot, "Lundin made the angle steeper than I wanted. It will make the focusing much more sensitive," he said. Slipher was apprehensive because Lundin had complained about the difficulty in figuring one of the disks, since it was so thin. E. C. Slipher was still in Phoenix.

103

The objective in its cell was bolted on to the upper end of the 5-foot-long (1.5-meter) tube. Now for the proof of the pudding. How good would the star images be and how large a usable field? Then Slipher had a "Beaverboard" shutter on hinges installed in front of the lens.

V. M. Slipher glued on a finely graduated scale along the slanted slot to permit a series of focal settings on a test plate. The first available evening, Slipher had me accompany him to the dome which housed the new telescope, where he pointed it to the star field in Orion (if I remember correctly). We went through a series of perhaps fifteen focal settings, which I recorded for him. After the plate was developed, he studied the rows of star images which had been spaced by a small change in the declination setting. Having determined the particular setting for the best focus the next day, Slipher loosened the clamp screw and turned the cell to the mark that corresponded to the best image.

Within a few nights, an 11-by-14-inch (27.9-by-35.6-centimeter) plate-holder was loaded, and Slipher, Lampland, and I went out to the dome for the first guided exposure.

V. M. Slipher set the telescope to the Belt and Sword region of Orion and had Lampland do the guiding for about half an hour. Lampland kept fussing, "You're not giving the telescope the best chance. We should have pointed the telescope nearer to zenith." But Slipher had his reason for doing so. He wanted to compare the plate to published photographs elsewhere.

After the plate was developed, fixed, and rinsed, he held up the plate, dripping wet, to the viewing light screen. We were all excited. It was beautiful. Then Slipher picked up the hand magnifier to examine the star images more critically. "They look pretty good," he said, "all the way across the plate. Think the telescope could use a 14-by-17-inch (35.6-by-43.2-centimeter) plate."

He passed the magnifier to Lampland, who inspected the plate for several minutes. Then they passed the magnifier to me. I was to do this on a thousand plates over the next fourteen years.

It is essential for one to inspect the star images in the central area, near each edge and end of the plate, and each of the corner areas. The purpose of this practice is threefold: to see if the plate had been seated properly in the plate-holder, since one must do this by 'feel' in the dark when loading; to see if the plate was bent to the proper curvature; and to see if the quality of the images was adequate when observing conditions were marginal. If the plate was not quite up to standard, it was necessary to know immediately in order to prepare the schedule for the following night of observing.

Dr. Lampland had to leave in another day or so for a spring semester lectureship at Princeton University. He was already several weeks late in departing. I could tell that he did not want to go. He had had a lot of experience in planet searches in earlier years and would have preferred to be involved in the launching of the new search with the new telescope.

In exchange, Dr. R. S. Dugan (pronounced "doo-gan"), his wife Annette, and two adopted children came for a semester of residence on Mars Hill. He intended to do a photometric study of certain variable stars, but his photometer would not fit the 24-inch (61-centimeter) refractor. So he spent the time reducing his old observations.

V. M. Slipher and I tested the new 13-inch (33-centimeter) telescope for several weeks. Several brands of 14-by-17-inch (35.6-by-43.2-centimeter) plates were tried to determine which to select for the search. A serious problem was encountered. The sagittal depth of the concavity of the photographic emulsion surface ranged or differed by a few millimeters. The stars would be well-focused in one area of the plate and seriously out of focus in another and cause a serious loss in magnitude. The field of good images on such an instrument is slightly concave toward the lens and is known as the Petzval curvature. The steeper the converging cone of light from each star (after passing through the lens), that is, the faster the focal ratio, the smaller the tolerance for departure from the Petzval curvature. For this telescope, the tolerance in focus was only .20 millimeters in a focal length of 1700 millimeters (or 66.5 inches).

Some brands of plates were flatter than others—too flat. For other plates the concavity almost fit the Petzval curvature, but would vary from plate to plate of the same brand. Unless the plates are made to conform to the proper curvature while being exposed in the telescope, the plates would not be sufficiently matched for examination under the Blink-Comparator later.

V. M. Slipher designed a plate-holder that took care of this problem beautifully. Three such plate-holders were then made in the observatory shop under his close supervision. Curved arc strips of brass were milled to serve as ridges at the midsides and midends of the holder. Small gusset faces were installed in each corner. On the removable back were five thumbscrews, one in the center and one in each corner. The plate was loaded emulsion-side down to rest on the brass ridges and the corner gussets. The plate-holder back was then laid on the plate and latched. If the plate was too flat, the thumbscrews at each corner forced contact with the corner gussets, making the plate more concave. If the plate happened to be too concave, the central thumbscrew was turned down to make the plate flatter.

In order to accurately determine the curvature of the plate, Slipher designed a testing table having a short, glass rod coming into contact with the emulsion surface from a hole in the center of the table. The other end of the glass rod was fastened to a long amplifying arm on a fulcrum to point to a linear scale where one could read the amount of curvature as one turned down the central thumbscrew. The corner screws had to be turned tight enough to prevent slipping of the plate in the plate-holder with changing hour angle at the telescope. If too tight, one would snap off the corner of the plate. After the curvature was adjusted, the lever was hooked up to remove the glass rod from the plate. Then the large slide was inserted to cover the plate from light. This testing, of course, had to be done in the dark, except for a faint red light by which we read the curvature scale.

Thereafter, every plate was adjusted for curvature before being placed in the telescope. The search would have been impossible without it.

This seemed to work very well. After several trial plates, where Slipher watched me set the telescope to the proper region by the setting circles, carefully guide the telescope, closing the dome at the end of the night's work, and so on, he seemed satisfied with my operation. In effect, he said, "As far as the work in the dome is concerned, you are on your own." Here I was, entrusted with an expensive instrument and charged with keeping it busy with productive work.

In the office, V. M. Slipher handed a Norton star atlas to me and showed me how to select the guide stars along the Ecliptic for the proper spacing of the plates. Then he gave me a new observer's record book where I would jot down the serial plate numbers, the right ascension and declination coordinates of the guide star, the dates and exposure

Clyde W. Tombaugh entering the 13-inch (33-centimeter) telescope dome, carrying one of the 14-by-17-inch (35.6-by-43.2-centimeter) plate-holders.

times, the hour angle, the condition of the sky, the steadiness of the guide-star image, and the temperature of the night air. This was to be the permanent plate record.

It was now April, and V. M. Slipher was anxious to get the plate-taking started. But there were still other vexing problems ahead. One was in the driving mechanism for tracking, as the telescope was slowly turned westward to counteract the Earth's rotation eastward. For regions at certain hour angles, the guide star would oscillate in an east-west direction over a small arc, but too great to be tolerated for obtaining point star images. It required a lively manipulation of the guiding control buttons to keep up with the guide star, otherwise the star images would be short trails. This sort of thing soon wears out the observer. If one adjusted the screw more tightly into the teeth of the large worm wheel, there was too much friction and the driving mechanism stalled. There had to be a slight amount of play in the engagement of the teeth.

I tried adjusting the counterweight on the opposite side of the polar axis from the telescope to balance the telescope more perfectly. Now this pulsing effect, as we called it, was worsened. The worm would give the teeth in the worm wheel a nudge then coast for a few seconds until it caught up with it again. It became obvious that one must keep the leading edge of the worm in continuous contact with one particular side of the teeth in the large worm wheel. Then I readjusted the counterweight so as to increase the resisting torque on the worm wheel, a little at a time. The driving improved. After a few more trials, the pulsing stopped, and the driving was steady. What a relief!

On 6 April 1929, I took plate number one, guided on star Delta Cancri. It was a good night with a light wind, temperature 22 degrees F. The search for a trans-Neptunian planet was renewed.

The plate regions were to be duplicated within a week, under the same conditions, in order to have a pair for comparison to detect the motion of such a body.

Dr. V. M. Slipher had selected the Cramer Hi Speed brand of plates as having the most suitable characteristics for the search. These plates were blue-sensitive, which held three advantages. Lowell had expected his Planet X to be bluish, like Neptune, and the 13-inch (33-centimeter) telescope had been figured for the best focus in blue light.

The plates were not yellow-sensitive, therefore they did not produce a star image with a blue core, surrounded by a halo disk with the out-of-focus yellow image. Finally, blue light is the most actinic in photography.

As we learned later, Pluto turned out to be yellowish, and only a fraction of its light was bluish, resulting in a photographic loss of nearly one magnitude. Thus, Pluto was only about half as bright as it would have been if it had been photographed in yellow light! It is virtually impossible to find a yellow-sensitive plate (isochromatic) that is not also sensitive to blue light. To set the focus for yellow light would have required a yellow filter in front of the plate to absorb the blue rays and eliminate the out-of-focus halo of blue light. This would have required a 14-by-17-inch (35.6-by-42.2-centimeter) gelatin filter which is quite vulnerable to damage and collection of dust. The photography was difficult enough without this sort of encumbrance, especially in a program that required so many plates to be taken in rapid succession.

The season was not the best one for observing. Frequently, insidious thin haze would creep over the sky. During each plate exposure, I would take my eye from the guiding eyepiece to glance up at the sky visible through the wide slit of the dome. Of course, if clouds drifted in, they could be easily detected by the areas of blanked out stars. If it was thin haze, it could only be detected by a mental integration of the stars visible compared to what ought to be visible. Dr. Slipher set the standard exposure at exactly one hour. Many such nights could be salvaged by prolonging the exposures by ten minutes, fifteen minutes, twenty minutes, or in some cases by a half-hour. This was necessary to be able to match the magnitude limit of the other plate. Unless the magnitude limit was well matched, a pair of plates was unblinkable, and the region would have to be done over.

It was not until many months later when I started blinking the plates regularly that I learned to prolong exposures correctly. The Blink-Comparator quickly reveals how much a given haze coverage affects the magnitude limit and how little tolerance there is for it.

The April lunation was a fair run. It was necessary to make repeat plates on all three plate regions in the constellation of Gemini because of casualties. In the easternmost region, guided on Delta Geminorum, I had to make four plates. These contained the images of Pluto, but we did not know it until ten months later.

The westernmost plate, 36 Geminorum, was right in the middle of the Milky Way. The number of star images was bewildering. After I developed the plates, Dr. Slipher had Mrs. Fox count the number of star images in a few half-inch squares in order to estimate the total number on the plate. The number: over 300,000. No wonder Pluto remained undetected in the earlier search, camouflaged against such a background of stars. On a single plate, Pluto's image would look just like the other thousands of faint star images. Slipher was aghast at the 300,000 images. So was I. Then I commented that I was glad I didn't have to search through that mess. Oh how I would eat those words later! The task of examination looked most formidable, indeed.

With the tenth plate, I had a new type of plate casualty. I think it may have been one of the Delta Geminorum plates. I was guiding away peacefully, when suddenly, "BOOM!" It was so loud that it lifted me right off of the observer's chair. *What was that?* I was petrified with fright. *Did the thin lens component break? How would I explain this to Dr. Slipher? Well, there wasn't any tinkling of glass falling down in the tube. The plate must have broken.* These thoughts went racing through my mind.

Some stars explode now and then, here and there in our galaxy. These stars are called *novae*. Normally, they are very hot subdwarf stars. Occasionally, one flares up a thousand or more times brighter within a day. Now such exploding stars at distances of thousands of light-years are not supposed to have this kind of effect in an astronomer's telescope.

Well, a little later, another plate broke in the same way, with a large "BOOM!" I don't see how a plate breaking could make so much noise, unless it is amplified in the tube like a drum.

I reported these incidents to Dr. Slipher with embarrassment. I asked him how much these large plates cost apiece. "A little over $5," he said. This was in 1929. Now they cost about $30. Well, Dr. Slipher didn't seem to know what to do about it either.

I could ill-afford to spend observing time repeating a plate in such a heavy schedule. Besides, the plate breaking was getting on my nerves. Like when the fellow in the room above you drops a heavy shoe on the floor, you wait for the other one. Only this was worse—not only was the noise much greater, but I was waiting all night for it to occur.

I had to find some way to stop it. Then I got to thinking. When I have to turn the central screw more than usual, the plate cannot spread out because the corner screws have the plate locked in place. If I loosened the corner screws a little, the plate could spread slightly as I turned down the central screw to achieve the proper curvature. Then I could lock the corner screws. That did it. I never had a plate break since. After experiencing no more breaks, I explained to Slipher how I stopped it. He was delighted.

There was yet another kind of plate casualty to plague me. When the exposure was carried through an hour angle point of 42 minutes west of the meridian, a little "chug" noise on the declination axis occurred. The guide star would suddenly be offset from the intersection of the illuminated threads by about 20 seconds of arc. Quickly, I got the star back to its proper place. The declination axis had slipped to the west about 0.5 millimeter. The continued exposure recorded all the stars as a separate set of images slightly to one side of the others. Every star was a double image, making the plate unusable for blinking. However, it could serve as a check plate for possible planet suspects.

I attempted to remedy this situation by tightening the large declination axis collars against the polar axis in order to eliminate the slight play. However, the chug could not be completely eliminated until the collars were so tight that I could not move the telescope in declination. Since this attempted remedy failed, the collars were loosened to their original position.

Thereafter, I initiated another precaution. If the schedule indicated that an exposure would be carried through the critical hour angle of 42 minutes west, I would swing the telescope west enough to chug the declination axis and bring it back to the proper right ascension before beginning the exposure. I had no more problems with double images after following this practice. As I recall, this was the last of the mechanical and optical difficulties encountered.

Every new telescope has its own set of idiosyncrasies. It just takes some time to discover them. Some of the faults can be remedied by a minor modification of the equipment, while others may be avoided by a change in operational procedure. The new 13-inch (33-centimeter) telescope with its longer focal length and much larger plates was much more sensitive in operation than the smaller cameras used in the past. The techniques of operation had to be much more refined.

The Gemini region is just northeast of Orion, which was sinking low in the evening western sky, over three months from opposition. This was certainly the wrong time to photograph Gemini, but Dr. Slipher wanted me to do so because this was one of the regions Lowell favored for his Planet X.

Lampland would write letters from Princeton to V. M. Slipher asking about the progress. In one of them, he inquired about the blinking of the plates. Had the modification of the Blink-Comparator been made to accommodate the much larger 14-by-17-inch (35.6-by-43.2-centimeter) plates? This was not easy. Slipher devised large, treated, wooden, ½-inch boards attached to the original metal panels that held the racks for the earlier smaller plates. The arm holding the right-angle prisms, the shutter, and eyepiece had to be set further out by a thick spacer to allow the large plates travel room under it. On the wooden boards were thumbscrew clamps to permit setting various sections of the large plates for accessible blink examination. It required ten sections to cover each pair of the large plates. Each change of section required readjustment in the horizontal, vertical, position angle, and radial depth by fine-threaded hand-controls. With four variables, each affected by the others, it took half an hour to get them all properly adjusted so that the fainter star images (.03-millimeter diameter) would not "wiggle" during the rapid alternating views throughout the 7-inch (17.8-centimeter) travel of the plate carriage (the half width of the plates). After a few weeks experience, one learned to make the necessary complex adjustments in about two minutes. This procedure must be repeated for each new section to be examined. At that time the observatory could not afford to build or buy a much larger Blink-Comparator, with which a large plate could be examined with one set of adjustments.

But who would scrutinize these pairs of plates teeming with myriads of star images?

After I had obtained the three pairs of plates in Gemini (Lowell's favored region), the two Sliphers started blinking them in an effort for a quick find of Planet X. They took turns working at it for several days. But one cannot make a thorough examination of these star-rich regions in Gemini without several weeks of work. They missed seeing the faint images of Pluto. After the discovery in the following year and then knowing just about where to look, the images of Pluto were found on those 1929 plates.

The Gemini plates were taken at a large angular distance from the opposition point (near 90 degrees). The asteroids had passed their post-opposition stationary point and their apparent motion was rapidly eastward instead of retrograde. Pluto was near its stationary point and would not have exhibited a detectable shift without a longer interval of time between the two plates of the pair. Moreover, the tangential vector would have been too uncertain to indicate that the body was beyond the orbit of Neptune. I have later wondered if this situation was fully taken into account in that May examination.

After that first blinking run, I sensed a feeling of frustration, almost one of despair. Dr. Lampland was the one most experienced in blinking work, and he was still in Princeton. It was not until sometime later that I realized how desperate the situation was at that time. During the preceding years, they had come to believe that the panacea to the failure of finding Planet X was a more powerful instrument. However, that was only part of the difficulty. It was the "quick find" attitude and procedure in the earlier searches that was defeating them. As an example, I witnessed for myself the hasty manner of the May blinking, in spite of now having the super instrument. Moreover, several important aspects of proper observational procedure were ignored, as I was to learn for myself a few months later.

As I look back to the year 1929, Dr. Slipher was in a predicament. They had all done a superb job in designing and constructing the excellent 13-inch (33-centimeter) telescope. Slipher had explained to the new trustee, Mr. Putnam, that such an instrument was essential to finding Lowell's Planet X. Now they had it. Slipher had stuck his neck out to Putnam. In turn, Putnam had stuck his neck out to his Uncle Lawrence Lowell in getting a large sum of money to build it. Naturally, Dr. Lawrence Lowell was interested in getting his brother's Planet X found. Putnam was doubly interested in finding his Uncle Percy's Planet X. For a while, Dr. Slipher probably felt that he had stuck his neck out hiring a green Kansas farm boy with only a high school education to do the work. But he didn't have much choice because the observatory budget was too strained to hire a highly trained person. I felt under some strain to measure up to Dr. Slipher's hopes and expectations.

As news of the renewed planet search at Flagstaff became known to other astronomers, many regarded it as foolishness. I got the direct impact of some of that feeling myself in 1929.

With the completion of this powerful new telescope, the observatory members felt that they had reached the promised land. It had cost a great amount of money, work, and time. Now with the much greater number of stars recorded on the plates, the grim task of blink examination, and the high expectation of positive results, the new instrument was beginning to look like a monster of greater liability. I can never forget how the first good plates taken of the Zodiac portion in the Milky Way overwhelmed Slipher. Later, in the blink examination, I came to dread the Milky Way regions.

Many a night I saw the Pleiads, rising through the mellow shade,
Glitter like a swarm of fire-flies tangled in a silver braid.

Tennyson, from Locksley Hall

10

The Photographic Search Begins

by Clyde W. Tombaugh

When the photographing began in April, my instructions from Dr. V. M. Slipher were, "Do the regions in Gemini and proceed eastward along the Ecliptic as rapidly as possible." There was no warning about the improper procedures and strategy.

The Gemini region was already about 90 degrees west of the opposition point. It was not until the end of the June lunation that I succeeded in catching up to the opposition point, which sweeps eastward through the constellations at a rate of 30 degrees each month. This is caused by the Earth's motion of revolution around the Sun.

On 7 June 1929, V. M. Slipher attached the 5-inch (12.7-centimeter) Cogshall camera on the side of the 13-inch (33-centimeter) telescope tube. The purpose was to serve for confirmation of planet suspects seen on the 13-inch (33-centimeter) plates. Half of the June lunation work was over. This was a beautiful camera with a field a little wider than the 13-inch (33-centimeter). The Cogshall camera took 8-by-10-inch (20.3-by-25.4-centimeter) plates. The focal length was almost exactly one-third that of the 13-inch (33-centimeter). Consequently, the

angular scale was one-third as much, namely: 1 degree was 1 centimeter across instead of 3. This means that to attain the same linear
shift, the interval of time between the plates of a pair must be three
times longer. This longer interval is not compatible with the ideal interval on the 13-inch (33-centimeter) pairs.

The 5-inch (12.7-centimeter) camera plates lacked two stellar
magnitudes of reaching as faint as the 13-inch (33-centimeter) plates.
Later, I was to find that about 90 percent of the planet suspects occurred in the faintest two magnitudes and could not be checked with
the 5-inch (12.7-centimeter) camera plates.

I rigged up a device so that the 5-inch (12.7-centimeter) camera
shutter opened and closed simultaneously with the 13-inch (33-centi-
meter) for beginning and ending the exposures. The Cogshall camera
yielded superb star images over a field nearly 20 degrees across and
recorded stars to the sixteenth magnitude. It was continually used with
the 13-inch (33-centimeter). Eventually, fourteen years later, the entire heavens visible from the latitude of Flagstaff was covered by the
5-inch (12.7-centimeter) camera plates. This served as a library of
plates for reconnaissance of other types of objects. However, it was of
little aid in the Planet X search, although I could have discovered
Pluto with it, but barely. For me, it meant loading and developing
more plates and writing more observational notes.

By 14 June, the Moon was in first-quarter phase, and I could take
no more photographs until after midnight when the Moon set. I was
now photographing in the constellations of Scorpius and Sagittarius, in
the regions directly toward the Galactic Center. The plates were beautiful—strewn with those fascinating globular star clusters, galactic
(open) star clusters, gaseous nebulae, and strange lanes and patches of
the 'dark nebulae' (cosmic dust clouds) revealed by their blanking out
the stars in those places. These plates contained three times the number of star images on the Gemini plates. It was fascinating. I thought,
Somebody is going to have a big job examining these plates.

By 17 June, there were no more moonless hours left of the night. I
was tired from loss of sleep, having taken a considerable number of
plates and developing them. There was some interesting astronomical
reading I wanted to do, particularly the papers by William H. Pickering in the current issues of *Popular Astronomy* about the several
planets he had predicted.

I had now taken about one hundred 14-by-17-inch (35.6-by-43.2-centimeter) plates, each of one-hour exposure. Only a few pairs had been blinked. Lampland had returned from Princeton, his desk piled with mail from several months absence. Nobody was doing any blinking of the plates.

V. M. Slipher was getting desperate. Mrs. Fox, the secretary, was leaving because her husband had gotten a job in California. Dr. Slipher was now confronted with finding a new secretary and training her in the unusual nature of duties in an observatory.

About 18 June, Dr. Slipher came to my office and said he wanted me to start blinking the dozens of pairs of plates that had accumulated. I was overwhelmed. I had thought that the three older men would go after the blinking. It had become evident to me that the one doing the blinking carried the heavy responsibility of finding, or not finding, the planet.

The summer rainy season was approaching when little photographic observing could be done because of clouded night skies. I would have plenty of time for blinking. Dr. Slipher intended to keep me busy.

I was a perfectionist. Nothing short of perfection would satisfy me. When I dug that cave pit on the farm, I shaved the sides with a spade like a sculptor. When I planted the kafir corn and milo maize, the rows across the field had to be straight as an arrow or I was unhappy. Later, every planet-suspect, no matter how faint, had to be checked out with the third plate—either yes or no, not maybe. I had always taken seriously one of the adages in one of my school books: "Do a thing well, or don't do it at all."

I proceeded to blink two pairs of my plates in a thorough manner. It was the most tedious work I had ever done. I encountered several dozen asteroids which shifted in position in the interval between the dates the plates were taken. How would I know which one was Planet X? Several of these plate regions had no third plate and some were taken near the asteroid stationary points. I just could not reconcile myself to conducting a planet search on such a hit-and-miss manner. I was in a state of despair. My morale was at its lowest ebb. With this uncertainty, I could see no point in going on. I fell to doing some soul-searching. Whatever other scientific endeavors that I might aspire to, I was blocked by the lack of a university education and degree.

To make matters worse, late in June 1929, an astronomer in his sixties from an eastern observatory paid us a visit. Dr. Slipher and I had taken him out to show him the new telescope and explain its program of work. No sooner had we returned to the administration building, when the phone rang for Dr. Slipher. Slipher went down the hall to his office to take the call. The visiting astronomer leaned over to me and said softly, "Young man, I am afraid you are wasting your time. If there were any more planets to be found, they would have been found long before this."

Feeling obliged to defend Dr. Slipher, I said, "We have one of the most powerful planet-search instruments in the world. I doubt if anyone has searched to such a faint limit. I am going to give it all that I have." In spite of my despair with my recent blinking, I wasn't going to admit defeat.

My office was just across the hall from Lampland's. He sensed that I was a little depressed. He sympathetically tried to cheer me up, saying, "I know what it is like to be homesick." It was the longest time I had ever been away from home. In July, the cloudy season set in, and I was allowed to take the long train ride home for a three-week, paid vacation. I would get to see my new baby sister, whom I had not seen.

There was still one week of the wheat harvest left, and so I was back to cutting wheat again. I didn't mind the dust and chaff; it was good to have a wheat crop to cut.

After three weeks, I boarded the Santa Fe train back to Flagstaff. I would not see the family again for nearly a year.

The cloudy season continued, and so I plunged into studying the asteroid problem and apparent motions of the planets. I recalled the different retrograde arcs of the planets as I had observed them in the sky in my earlier years. Now I would inquire into the subject with more quantitative exactness.

From the *American Ephemeris and Nautical Almanac*, I studied the daily coordinate positions of the planets Mars, Jupiter, Saturn, Uranus, and Neptune over a span of three years. I noted the changing rates of apparent motion, especially in relation to the opposition positions and their various apparent stationary points, the arc-spans, and times of apparent retrograde motion. I saw the solution to my problem. A simple extrapolation indicated what to look for in identifying

Planet X and how to avoid confusion with the hundreds of asteroids that I would encounter.

I was now filled with hope and enthusiasm. Now it appeared possible to make a thorough, systematic search down to the seventeenth magnitude. I resolved now to bite the bullet and go through with it. Every planet suspect encountered to the limit of the plates could be checked off on the third plate with a definite verdict. With the plates taken at or near opposition, I would be able to determine immediately the distance of a planet suspect because the shift in position was an effect of parallax.

I had only a fragmentary knowledge of the earlier search work. Because of the uncertainties of theoretical predictions, I would search through the entire Zodiac. I might find several planets, whether predicted or not predicted, with the new, very effective equipment in hand.

The summer rainy season at Flagstaff causes a scramble with respect to keeping up with the opposition point. Some years, one is totally shut out for two months. Other years one may manage to get a few plate regions in that area. In the June lunation, one photographs as far east of opposition as one dares. Then in September, hopefully, one can resume the strip before the unphotographed area is too far west of the opposition point. Of course, one must calculate the necessary overlap, since any planet in the region has been moving in a retrograde direction (to the west) for two months. In June 1929, I was unable to do this. Now in September, the constellation of Capricornus was too far west of opposition. This break would have to be retrieved next year.

As the skies cleared in September, I started photographing in the constellations of Aquarius and Pisces. These were lovely regions with hundreds of spiral galaxies to view and not so populous in stars, only about 50,000 per plate. The regions were 60 degrees from the equator of the Milky Way. I could blink a pair of plates in three days of work.

I knew that Uranus was in Pisces, but I did not want to know exactly from its listed position in the *American Ephemeris*. I wanted to test myself on the surprise of encounter.

When the Moon stopped my night work, I started blinking these new plates in daytime, field by field, strip by strip, panel by panel. Suddenly, upon turning to the next eyepiece field (1-by-2-centimeters),

Clyde W. Tombaugh at the guiding eyepiece of the 13-inch (33-centimeter) telescope.
Taken in 1931 (one year after the discovery of Pluto), at the age of 25 years.

there was Uranus. Being of the sixth magnitude, it was a real wallop. I almost ducked my head. I stopped and measured its shift in position with a millimeter scale. It was exactly as I had calculated it would be on those plates. Now my confidence was complete. From now on, for the next several years, I was obsessed with the planet search. Would I find a trans-Neptunian planet, or not? Considering each star image as a straw, I would sift through every straw in the haystack to find the needle. By doing the blink examination very thoroughly and if nothing was found, I would be able to state that such a planet does not exist.

After measuring the shift of Uranus, I called in Lampland and the Sliphers for a blink view of Uranus for the first time with the new 13-inch (33-centimeter) telescope.

Later that fall, E. C. Slipher got to reminiscing about some of the earlier search work. He dug out an early plate and pointed to an image of about the thirteenth magnitude with a "?" marked on the back with ink. "This might be Planet X," he said. I was stunned for a moment, and then I blurted out, "You mean that you don't know?" Perhaps there had been some misfortune on confirmation, I don't know. Perhaps he intended to fire up my enthusiasm. He did. The thought flashed across my mind that my chances of finding big game were better than I had at first anticipated.

Both of the Sliphers had grown up on a farm in Indiana and then earned degrees from Indiana University. Perhaps this was a factor in V. M. Slipher's inviting me to come to Flagstaff. In those days on the farm, one learned to work hard. Well, they got what they wanted. I got four or five pairs of plates blinked in that light of the Moon period. The plates were well-matched, all duplicated at about the same hour angle and taken not too far from the meridian. I was learning how to schedule the photographing. I started counting and tabulating the number of galaxies visible on each of the ten panel sections of each pair of plates. (As a result of this study of the apparent distribution of galaxies over the following decade, I had gathered ammunition to challenge Edwin Powell Hubble on the question of the space distribution of galaxies.)

The following lunation, I moved eastward into Aries. While blinking those plates, V. M. Slipher would come down to the room where I was blinking. "Are you finding anything [planets]?" "Not yet," I said, "but finding lots of asteroids and variable stars, also some interesting groups of nebulae [now called galaxies]."

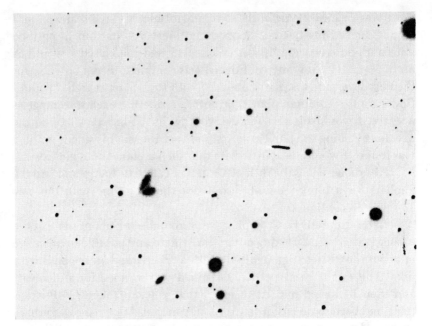

A small planet search field containing the images of a close pair of very distant galaxies and an asteroid trail. The asteroid traveled the length of the trail during an exposure time of two-and-a-half hours. This photo was enlarged from the original plate.

There seemed to be an increasing sense of anticipation. Perhaps Slipher was getting inquiries from Putnam. Likewise, Lampland seemed to be getting anxious. I worked furiously, totally abandoning the eight-hour day. I was feeling more and more a feeling of acceptance, as I was included on all the doings and discussions with occasional visits from other astronomers. Many stopped off at Flagstaff for a day or two on their way to or from the astronomical mecca, the Mount Wilson Observatory. Their 100-inch (254-centimeter) telescope was the largest in the world. I got to meet and talk with the famous astronomers I had read about in the books. I forgot all about homesickness. I was living it up. It was fascinating to hear what they were learning about the Universe in their researches.

Each succeeding month, I was getting closer and closer to the constellation of Gemini. It was also the region where Herschel discovered Uranus in 1781.

In November, I was going through the star fields in western

Taurus. The stars thickened in number. The search was approaching the Milky Way. By the end of November, I had blinked about ten pairs of plates. The last two contained as many stars as all of the other eight together. I had sifted through about one million stars.

The next blinking period in the middle of December was very difficult because there were so many stars. As I remember, only two pairs were blinked. No longer could I keep up with the examination of the plates. It was like plunging into a jungle.

The 13-inch (33-centimeter) plates were recording stars having the same intrinsic luminosity as the Sun to a distance of 6000 light-years. A spherical volume of space with a radius of 6000 light-years embraces an enormous volume of space to contain stars. Stars with intrinsic luminosities 100 times that of the Sun were visible on the plates to distances of 60,000 light-years. The 13-inch (33-centimeter) recorded images of galaxies to distances of several hundred million light-years. I was getting a real tour of the Universe. Although the sky appears as a two-dimensional surface, in reality one is seeing objects through an enormous range in depth. One searches for planets in the foreground projected against this vast background.

As one approaches the Milky Way, the number of galaxies visible drops off. In central and eastern Taurus (also in western Gemini) the spiral and elliptical galaxies are totally absent on the plates. This is caused by the thin interstellar dust clouds. They are hundreds of light-years in thickness and are opaque enough to hide not only the far-distant galaxies in the line of sight, but also the more distant stars in our own galaxy.

Taken under more optimum conditions, the plates in eastern Taurus and western Gemini were running about 400,000 stars per plate.

In the January 1930 lunation, I rephotographed the entire Gemini region. On 21 January 1930, I set the 13-inch (33-centimeter) telescope on Delta Geminorum again. A good night seemed to be in prospect. The sky was very clear. Within ten minutes after the shutter was opened to begin the exposure, a strong northeast wind sprang up. In another ten minutes, it was a howling gale. The guide star, Delta Gem, began to fuzz up badly into a diffuse patch, making it hard to guide. Then terrific gusts swept up the east side of Mars Hill. The star image would swell up to several apparent diameters of Jupiter. With succeeding gusts, the image would swell up so badly that it became invisible. I muttered, "I can't see anything to guide on." It was a most helpless situation.

After the gust had passed, the guide star became visible again, but it was still swollen and in violent, agitated motion. I had never seen such terrible seeing, nor have I seen such in the years since. I thought of terminating the exposure and closing the dome. I was getting worried about the gusts snapping the ropes that held the doors of the slit open, although the ropes were strong and new. With that kind of seeing, the plate was spoiled anyway but I decided to finish out the exposure just to see how bad the images would be after I developed the plate. I had more plates to take, but to do so would be futile, so I closed the dome.

After developing the plate the next day, I viewed the dripping plate with a magnifier. The images were swollen to several times their normal diameter. Also, the dilution resulted in a loss of about 1.5 magnitudes. Nevertheless, that horrible plate did record the image of Pluto, but I did not know it then. Some astronomical textbooks state that Pluto was discovered on 21 January 1930. Nonsense! Only on the date that the images are recognized as those of a planet does it constitute a discovery. I have written many letters to authors (many astronomers) to get them to correct the error for their next book edition.

On 23 January, I photographed the Delta Gem region again. It was a good plate. I was unable to photograph the region again until 29 January. This six-day interval was twice as long as I preferred. This was the pair that I would blink later.

During that lunation, I photographed the western regions of Gemini again.

When the Moon cut me off about 7 February, I resumed blinking the eastern Taurus plates. After those were finished, I could stand no more plates so rich in stars. I decided to skip over the star-rich plates of western Gemini. The Delta Gem plates were much less populous of stars, only about a thousand stars per square degree. I could probably finish that pair before starting the next two-week run at the telescope. My life was tightly regulated by the phases of the Moon.

There was also another reason for selecting Delta Gem for blinking. The bad 21 January plate was inadequate for checking possible faint planet suspects. If I was able to get through this region without needing a better check plate, I could leave it off the schedule in the next lunation's run at the telescope.

Then I felt like some watcher of the skies
When a new planet swims into his ken.

Keats

11

The Ninth Planet
Discovered

by Clyde W. Tombaugh

On the morning of 18 February, I placed the 23 January and 29 January Gem plates on the Blink-Comparator, starting on the eastern half. This was a most fortunate decision. Had it been otherwise, Pluto might not have been discovered in 1930.

By four o'clock that afternoon, mountain standard time, I had covered one-fourth of the pair. After completing a horizontal strip on the left half, I rolled the horizontal carriage back to the center north-south line (which I always drew as a thin ink line on the back of the later plate of the pair). I had established this habit of progressing to the left so that I would not forget which way I was going in case of interruptions.

I could vary the speed of the shutter as it switched views from one plate to the other by means of a rheostat. I found from experience that about three alternating views per second was the most efficient.

Clyde W. Tombaugh examining a pair of planet search plates on the Blink-Microscope-Comparator. Taken in 1938, this picture shows Tombaugh at the age of thirty-two. He spent 7000 hours examining 338 pairs of 14-by-17-inch (35.6-by-43.2-centimeter) planet search plates, seeing about 90 million individual star images.

I raised the eyepiece assembly to the next horizontal strip. At the center line, I had the guide star Delta Gem in the small rectangular field of the eyepiece. After scanning a few fields to the left, I turned the next field into view. Suddenly I spied a fifteenth magnitude image popping out and disappearing in the rapidly alternating views. Then I spied another image doing the same thing about 3 millimeters (or .125 inches) to the left. "That's it," I exclaimed to myself. Now which image belonged to which plate? I turned off the automatic blinker and turned the shutter back and forth by a small finger lever. The right-hand image was on the earlier plate (23 January). West was to the left in the field. Then I turned the shutter to view the 29 January plate. This image was to the left of the other. Retrograde motion alright! If the direction of shift had been to the east, then the images would have been either spurious, or were those of two independent eclipsing variable stars which happened to be caught in alternate phases of variation. Considering the interval between the plates, the parallactic shift indicated that the object was far beyond the orbit of Neptune, perhaps a thousand million miles beyond.

A terrific thrill came over me. I switched the shutter back and forth, studying the images. Oh! I had better look at my watch and note the time. This would be a historic discovery. Estimating my delay at about three minutes, it would place the moment of discovery very close to four o'clock.

For the next forty-five minutes or so, I was in the most excited state of mind in my life. I had to check further to be absolutely sure. I measured the shift with a metric rule to be 3.5 millimeters. Then I replaced one of the plates with the 21 January plate. Almost instantly I found the image 1.2 millimeters east of the 23 January position, perfectly consistent with the shift on the six-day interval of the discovery pair. The planet images probably were recorded on the three 5-inch (12.7-centimeter) Cogshall camera plates of those respective dates. I checked those with a hand magnifier. There were the faint images, all exactly in the same respective places among the small configurations of stars. Now I felt 100 percent sure.

I removed the poor 21 January plate, replacing it with the discovery plate and called to Lampland from across the hall that I had found a trans-Neptunian planet. He said, "I heard the clicking of the comparator suddenly stopped, then a long silence." He suspected that I had run into something unusual. I explained that the interval was six days, which plate was the later date, and that south was at the top in the field so that he could see that the motion was retrograde. He started studying the images.

Then I walked down the hall to V. M. Slipher's office. Trying to control myself, I stepped into his office as nonchalantly as possible. He looked up from his desk work. "Dr. Slipher, I have found your Planet X." I had never come to report a mistaken planet suspect. He rose right up from his chair with an expression on his face of both elation and reservation. I said, "I'll show you the evidence."

He immediately hurried down the hall to the comparator room. I had to step lively to keep up with him. Lampland said something like, "Looks pretty good," and surrendered the comparator to V. M. Slipher to take a look. E. C. Slipher was out of town for a few days.

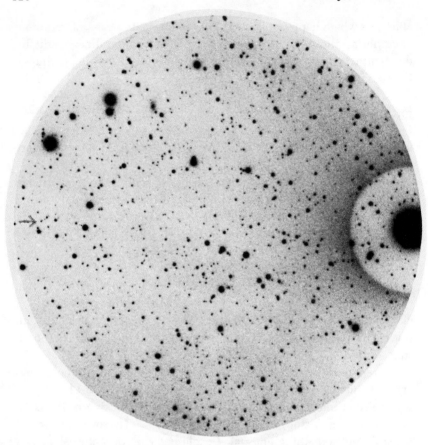

A very small portion, the size of a nickel, enlarged six times from one of the Pluto discovery pair of large plates. This one was taken 23 January 1930. The position of Pluto's image is indicated by a small black arrow in a sky area of 900 stars.

I explained that I had measured the shift to be consistent on the three plates, and that all of the images were in the correct positions on the Cogshall plates. Slipher kept flicking the shutter back and forth, studying the images. Then I said, "The shift in my opinion indicates that the object is well beyond the orbit of Neptune." I showed them the images on the other plates.

Then Slipher said, "Don't tell anyone until we follow it for a few weeks. This could be very hot news." Then he said to me, "Rephotograph the region as soon as possible." It had been a cloudy day, with

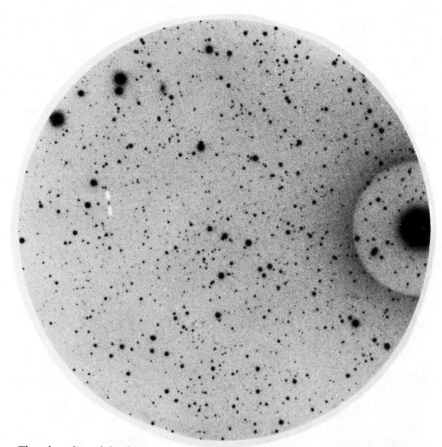

The other plate of the discovery pair. This one was taken 29 January 1930. Two white bands indicate the position of Pluto's image. The shift in position of Pluto was ⅛ inch, or 3.5 millimeters, on the original plates. This shift immediately indicated that the object was beyond the orbit of Neptune. The pictures are much enlarged. This field represents one part in forty thousand of the sky area blinked in the search of planets.

snow on the ground. I looked out the window, "Doesn't look very good for tonight."

The excitement was intense. Another era for the Lowell Observatory was suddenly ushered in. The announcement three weeks later would cause excitement all over the world.

At six o'clock, an hour later than my usual departure, I went downtown to pick up the observatory mail at the post office and eat my dinner in a cafe. After dinner, I looked up at the sky. No stars in sight. It looked hopelessly cloudy. I was so excited, yet I could tell none of my acquaintances. So I went to the only theater. It was Gary Cooper in *The Virginian*. With the gun duel, my knees were shaking worse than ever. After the show, I looked up: "Still cloudy." I drove up Mars Hill, put the car in the garage, walked over to the administration building, and sorted the mail into its respective slots. I tried reading. It was futile. I kept going out to look at the sky, with plate-holders loaded. By two o'clock in the morning, the sky was still cloudy. The Delta Gem region would be too far west and the Moon would have risen. So I went to bed. There would never be another day like that one.

The skies cleared off the next day and I prepared to rephotograph the Delta Gem region a little east of the meridian before the Moon came up. I developed the plate and put it to dry. After the plate was dry, I placed it on the Blink-Comparator to compare it with one of the plates of the discovery pair. Within a minute or two, I retrieved the planet image. After three weeks of no record, the planet was exactly where it should have been, having shifted westward by 11 millimeters since the 29 January plate.

I made a film contact of the central area containing the planet image. On the evening of 20 February, V. M. Slipher, Lampland, and I walked up the knoll to the 24-inch (61-centimeter) telescope dome. Slipher set the telescope to the region containing the planet, with the aid of the circles. Then he identified the immediate area with the film contact. Would so small a planet show a disk? This was uppermost in all of our minds. "Ah, there is the planet, slightly west of its position on the 19 February film." Pluto was viewed visually for the first time by the eye of man. Disappointingly, no disk was visible under high magnifying power.

The next clear night, Lampland started a photographic run on 4-by-5-inch (10.2-by-12.7-centimeter) plates with the 42-inch (106.7-centimeter) reflector at the Newtonian focus. After I made a few more large plates of the Pluto region, I started my photographic run with the 13-inch (33-centimeter) in the constellations of Cancer and Leo. It seemed best to continue photographing the Zodiac to completion. I suspended blinking, thinking the search was over—so I thought. I was getting tired of the tedious blinking, anyway.

E. C. Slipher returned and started visual experiments with a box having different size holes at various light levels. He placed it 1 mile (1.5 kilometers) east of the observatory and started observing the holes with the 24-inch (61-centimeter) refractor. Almost immediately, there was concern about Pluto's mass. The purpose of E. C. Slipher's experiment was to determine whether Pluto might be a larger body, but it failed to exhibit a perceptible disk because of the very low level of illumination by the Sun. E. C. Slipher's results indicated that Pluto could have a disk of .5 arc-second and still escape detection as a disk with the powerful 24-inch (61-centimeter) refractor, which is capable of revealing a brighter illuminated disk as small as 0.25 arc-second.

In the meantime, Lampland was taking several one-hour exposure plates with the 42-inch (106.7-centimeter) reflector at the Newtonian focus to search for possible Pluto satellites. Also, he took six- or seven-minute exposure plates on every possible night for the purpose of positions. The much longer focal length of the 42-inch (106.7-centimeter) reflector permitted more precise measurements.

Lampland also required the use of the micrometer eyepiece on the Blink-Comparator, which tied it up from further blinking for over two months.

Putnam, the trustee, was quickly notified of the discovery, as well as a few other close friends in confidence who had a special interest. I did not even hint of the discovery in my letters to my parents.

The discovery would involve considerable urgent correspondence for V. M. Slipher regarding impending questions. Other preparations were being made to brace them for the tide of problems that would strike following the public announcement of the discovery. Being the traditional "bad boy" in astronomy, the Lowell Observatory would face some unusual and difficult problems.

From night to night, Pluto continued to be on course perfectly. The time had come to make the plunge. I was in the secretary's office with V. M. Slipher on the evening of 12 March when he sent the announcement telegram to Mr. Putnam. He was to forward it to Harvard Observatory the next day for distribution to observatories and astronomers. This, in turn, would automatically be sent to Copenhagen, Denmark, for distribution in the eastern hemisphere.

The historic telegram read:

Systematic search begun years ago supplementing Lowell's investigations for Trans-Neptunian Planet has revealed object which since seven weeks has in rate of motion and path consistently conformed to Trans-Neptunian body at approximate distance he assigned. Fifteenth magnitude. Position March twelve days, three hours GMT was seven seconds of time west from Delta Geminorum, agreeing with Lowell's predicted longitude.

The announcement date, 13 March, was chosen to coincide with the 149th anniversary of the discovery of Uranus by Herschel and the 75th anniversary of Percival Lowell's birth.

During the previous week, Dr. Slipher had prepared a special Observation Circular, entitled "The Discovery of a Solar System Body Apparently Trans-Neptunian." It contained more detailed information on the discovery. Hundreds of copies were printed. The new secretary, Miss Constance Brown, and I ran the addressograph on the mailing envelopes, folded and enclosed the circulars, and put them in a box for the day of mailing. On 13 March, these circulars were mailed to observatories and astronomy departments all over the United States and the world.

After quoting the brief announcement telegram, the circular went on to describe the discovery in more detail.

The finding of this object was a direct result of the search program set going in 1905 by Dr. Lowell in connection with his theoretical work on the dynamical evidence of a planet beyond Neptune. (See L. O. Memoirs, Vol. I, No. 1, "A Trans-Neptunian Planet," 1914.) The earlier searching work, laborious and uncertain because of the less efficient instrumental means, could be resumed much more effectively early last year with the very efficient new Lawrence Lowell telescope specially designed for this particular problem. Some weeks ago, on plates he made with this instrument, Mr. C. W. Tombaugh, assistant on the staff, using the Blink-Comparator, found a very exceptional object, which since has been studied carefully. It has been photographed regularly by Astronomer Lampland with the 42-inch reflector, and also observed visually by Astronomer E. C. Slipher and the writer with the large refractor.

The new object was first recorded on the search plates of January 21 (1930), 23rd, and 29th, and since February 19 it has been followed closely. Besides the numerous plates of it with the new photographic telescope, the object has been recorded on more than a score of plates with the large reflector, by Lampland, who is measuring both series of plates for positions of the object. Its rate of motion he has measured for the available material at intervals between observations with results that appear to place the object outside Neptune's orbit at an indicated distance of about 40 to 43 astronomical units. During the period of more than 7 weeks the object has remained close to the ecliptic; the while it has passed from 12 days after opposition point to within about 20 days of its stationary point. Its rate of retrogression, March 10 to 11, was about 30" per day. In its apparent path and in its rate of motion it conforms closely to the expected behavior of a Trans-Neptunian body, at about Lowell's predicted distance. There has not been opportunity yet to complete measurements and accurate reductions of positions of the object requisite for use in the computation of the orbit, but it is realized that the orbital elements are much to be desired and this important work is in hand.

In brightness the object is only about 15th magnitude. Examination of it in the large refractor—but without very good seeing conditions—has not revealed certain indication of a planetary disk. Neither in brightness nor apparent size is the object comparable with Neptune. Preliminary attempts at comparative color tests photographically with large reflector and visually with refractor indicate it does not have the blue color of Neptune and Uranus, but hint rather that its color is yellowish, more like the inner planets. Such indications as we have of the object suggest low albedo and high density. Thus far our knowledge of it is based largely upon its observed path and its determined rates of motion. These with its position and distance appear to fit only those of an object beyond Neptune, and one apparently fulfilling Lowell's theoretical findings.

While it is thus too early to say much about this remarkable object and much caution and concern are felt—because of the necessary interpretations involved—in announcing its discovery before its status is fully demonstrated; yet it has appeared a clear duty to science to make its existence known in time to permit other astronomers to observe it while in favorable position before it falls too low in the evening sky for effective observation.
—V. M. Slipher

Flagstaff, Arizona
March 13, 1930

Immediately, the news of the discovery went over all the country by the Associated Press.

The editor of the Pawnee County weekly newspaper, *The Tiller and Toiler*, at Larned, Kansas, Mr. Leslie Wallace, phoned my parents in Burdett (on the edge of the county, 30 miles (48.3 kilometers) west). He said, "Did you know that your son discovered a planet? It is in the news everywhere." My father had answered the phone. Overwhelmed for a moment, he replied, "No, but I knew he was on the trail of one."

My mother was outside, perhaps hanging up the wash on the clothesline. He rushed out of the house. "Who was on the phone?" she asked. "It's about Clyde." Before he could continue, she asked, "Is something wrong with Clyde?" "No, Leslie Wallace said that the wires were reporting Clyde had discovered a planet! We must send him a telegram and congratulate him." The next day I received the telegram.

Realizing that I was young and inexperienced, V. M. Slipher kindly cautioned me to beware of getting involved with outsiders who had only their own greedy purposes. There were greedy wolves out there, packs of them. Right he was.

Telegrams poured in, congratulations, demands for exact positions, demands for exclusive stories, and so on. The Flagstaff telegraph office must have felt like a tornado had hit them. In a few more days, letters began to pour in by the hundreds. It was incredible! Story and feature writers and photographers swarmed over Mars Hill. The observatory secretary was frantically trying to attend to her duties. The number of daily observatory visitors increased greatly. The following summer, there were hundreds of visitors daily when people were making vacation trips. They wanted to see the place where Pluto was discovered.

From day to day, Dr. Slipher would inform the rest of the staff of some of the things going on. Every few days, there were discussions as to what policy to pursue on difficult problems coming up. The pressure those three older men were under was little short of outright trauma. There was danger of larger institutions snatching unearned prestige for themselves and putting the Lowell Observatory in the shade. The vindictiveness against Lowell flared up again.

Putnam kept pressuring Slipher to select a name for the new planet before someone else did. This privilege really belonged to the Lowell Observatory. There were outside political pressures on naming the planet. Indeed, I received a letter from a young couple in another state, asking that the new planet be named after their newborn child. At first, Mrs. Lowell proposed the name of "Zeus." Then later, she wanted the planet named "Lowell." Still later she wanted it to be "Constance," her own given name. No one favored that name. It was a touchy situation.

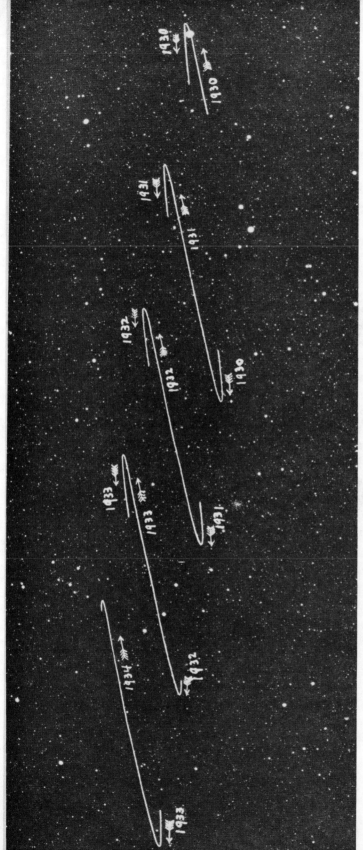

The apparent path of Pluto through the eastern portion of the constellation of Gemini, plotted by C. O. Lampland from positions on some one hundred plates. The disconnecting gaps are due to Pluto being behind the Sun during the summer seasons. Opposition time occurs in the middle of the long, retrograde traverse midway between the two apparent "stationary" points, where the planet reverses direction of apparent motion. Pluto makes 248 such loops during one revolution around the Sun, equal to 248 Earth-years.

In the meantime, over a thousand letters poured in, including those from several other astronomers, suggesting names for the new planet. Three names about equally headed the lists: Minerva, Pluto, and Cronus. It is customary to name planets after mythological deities. If Minerva, the goddess of wisdom, had not already been given to one of the asteroids, the name of the new planet would have been Minerva. Had not Cronus been proposed by a certain detested egocentric astronomer, that name might have been considered. Pluto, the Greek god of the Lower World, seemed the best one to pick. Outside of the Lowell staff, the name Pluto was first suggested by Miss Venetia Burney, age eleven, of Oxford, England. It was cabled by Professor H. H. Turner.

Remembering that Uranus went through three name changes, we wanted to select a name that would stick. Accordingly, the name Pluto was proposed to the American Astronomical Society and the Royal Astronomical Society of England. Both of these bodies approved the name unanimously. By taking the first two letters, the planetary symbol became "♇," for Percival Lowell. Years later, as a result of the naming of Pluto, the well-known fissionable element was named "Plutonium."

Putnam came to Flagstaff in April to assist and discuss many matters. He and I would drive downtown together for breakfast in a cafe. I had the greatest admiration for him.

In the ensuing months and years, many pertinent questions arose about the theoretical predictions and nature of Pluto, sparking intense controversies among many astronomers.

12

Problems of Pluto

by Clyde W. Tombaugh

The discovery of the new planet in 1930 caused a tremendous sensation, both among the general public and among many professional astronomers.

Pluto's unexpected faintness by nearly three magnitudes implied insufficient mass to be Lowell's Planet X. Attempts to find a satellite failed and thus a reliable means for determination of the mass was denied.

Pluto's light was yellowish, not bluish like Neptune. No disk was visible, implying a much smaller diameter. Instead of being the large, gaseous type of planet that Lowell expected, Pluto seemed to resemble the four inner planets (terrestrial-type), with higher densities and silica composition.

Everybody was interested in an orbit for more clues on this strange body. Several astronomers wanted to be the first to compute the orbit.

Immediately after the announcement from the Harvard Observatory news distribution center, Dr. Harlow Shapley, the director, was besieged with requests for positions by which an orbit can be computed. Shapley had no positions to give. Only the Lowell Observatory had the earliest positions, from the discovery plates taken in late January and plates taken from 19 February to 12 March, a span of only forty-nine days. Shapley pleaded with Slipher to give some positions and relieve the pressure.

Slipher wanted to keep the prestige of the first orbit for the Lowell Observatory. He held the trump card of the January positions, which gave the longest interval available. For the others, the earliest position available was what they could observe and measure following the announcement of 13 March.

One of the first telegrams to Flagstaff requesting planet positions was from A. O. Leuschner of the Students Observatory of the University of California at Berkeley. They were specialists in computing the orbital elements of asteroids and comets. Slipher was well aware of their experience in this field. Leuschner gave not a word of congratulations. I remember how annoyed Slipher was about this.

"All Leuschner is interested in is what he can get out of it," he said. "After all the long, hard search we did to find the planet." So Slipher ignored him.

About a week later, a second request came from Leuschner, this time profuse with congratulations. This did not impress Slipher, because Leuschner had made a statement to the press which seemed to cast doubt on the trans-Neptunian nature of the new object. Slipher was getting irritated with others who were grabbing off a lot of unearned publicity with premature statements. Leuschner thought that an orbit would have to be calculated to establish that the new object was trans-Neptunian.

Here indeed was an example of a professional astronomer, like many others, who did not understand the essence of the observational strategy involved in detecting a planet and immediately ascertaining its approximate distance. When the plates are taken at opposition, it is simply a matter of parallax. With a daily shift that small, it could be nothing else than trans-Neptunian. Then the manner of Pluto slowing down after opposition as it approached its apparent stationary point was further proof.

Of course, this does not tell what kind of an orbit the new body has. Whatever kind of orbit, Pluto was at some point in its orbit in 1930 that was far beyond Neptune's orbit. If the orbit is not too elliptical, one could assume a circular orbit with a radius equal to the distance indicated by the opposition parallax. As it turned out, Pluto was near its mean distance from the Sun in the year of its discovery in 1930. By using Kepler's Third Law, one could compute approximately where to look for it on old plates taken decades earlier. This is what other astronomers wanted to do. With a longer time span, a sufficient arc of the orbit would permit the determination of a reliable orbit.

Until now, the program of work at Lowell Observatory did not involve orbit computation. Now they suddenly needed it. Slipher knew that others would soon be computing an orbit. A little earlier, Putnam had been pleading with Slipher to release the earlier positions, lest a feeling of resentment be created among other astronomers.

The senior Lowell staff had taken a course in orbit computation when they were students, but after many years much had slipped away. So Slipher immediately sought the help of their old teacher who kept in trim by teaching the course year after year. This was Dr. John A. Miller, who was now at Swarthmore. Miller agreed to come to Flagstaff to guide the computation.

Dr. Miller arrived in early April, and the orbit computation immediately got underway. After fifty years, I can still see Dr. Miller, the two Sliphers, and Lampland seated at a long table furiously laboring on the computation with six-place logarithm tables.

Four days later, on 12 April, as the elements came out at the end, the orbit was shocking. Pluto's orbit was enormously elongated, with an eccentricity of 0.9 and a period of some 3000 years. The semimajor axis came out as 217.5 astronomical units, which means that Pluto would recede to a distance of 395 astronomical units at its aphelion. The orbital inclination came out as 17 degrees 21 minutes.

Immediately, someone of the elder staff made a three-dimensional model of this orbit. I was terribly dismayed that the new-found planet was so unorthodox in nature. The longitude of perihelion came out as 12 degrees, 52 minutes east of its ascending node. Since Pluto was almost on the Ecliptic at the time of discovery, at its ascending node of 109 degrees, 21 minutes, it implied that we had caught the planet slightly before its perihelion.

Well, one simply cannot get reliable values for the semimajor axis and eccentricity from positions over such an extremely small arc of its orbit.

These strange orbital elements were immediately published in a second Lowell Observatory Observation Circular. Also, V. M. Slipher stated that the new body may be comparable to Mars in size and mass. It would outshine the largest asteroid, Ceres, whose diameter was thought to be 480 miles (773 kilometers), by a hundredfold, or 5 magnitudes if Ceres were removed to the distance of the new object.

A quick reaction followed. Several astronomers began calling the new body a trans-Neptunian comet.

In the meantime, Leuschner and two of his graduate students, Ernest C. Bower and Fred L. Whipple, attempted to compute an orbit from positions spanning only nineteen days, from 16 March to 4 April. They computed several orbits, two of them parabolic, which is fairly typical of comets. These were announced and described in a Harvard Announcement Card dated 7 April and a week later in a Lick Observatory Bulletin. However, the computations did give a value of distance from the Earth of 41 astronomical units, a longitude of ascending node of 109 degrees, and an inclination of 17 degrees—all close to the actual values. But the elements of semimajor axis and eccentricity were about as wild as the Lowell values.

This object certainly did not resemble Percival Lowell's predicted Planet X. Newspaper headlines started flashing again on the seemingly extraordinary character of the new-found body. Instead of lessening the interest, it greatly enhanced the interest. Shapley got very excited about it. He and Leuschner both proposed that the new body was indicative of an entirely new class of Solar System objects. Were there more such objects roaming around in the outer reaches of the Solar System? This was causing considerable concern at Flagstaff. Putnam tried to assure us that whatever the nature the object might turn out to be, it was still a great discovery.

No one had ever observed a comet beyond the orbit of Saturn. Even Halley's comet faded to a faint, hazy spot on photographs taken with large telescopes when it receded to the distance of Jupiter. Yet we know from the computed orbital elements that Halley's comet recedes to a little beyond that of Neptune's orbit, but not as far as Pluto's mean distance, before it turns around and returns to the Earth's neighborhood thirty-eight years later. Hence, if the new object was a comet, it would have been a super-super comet. This was bound to create an even bigger sensation, both to the general public and also to astronomers. The new-found body had simply exceeded all precedents by a wide margin.

As E. C. Slipher argued, Pluto's image was strictly starlike in the 24-inch (61-centimeter) Lowell refractor, with no hint of a coma head, suggesting a comet. However, as we know now, at Pluto's distance from the Sun, the temperature would be so low as to probably freeze all the volatile substances to a solid nucleus, with no visible coma.

Some tried to call the new object a large asteroid. Well, if all the hundreds of asteroids were rolled together into one, it would not equal the probable mass of the new object.

After the discovery was announced, it seems that only one outside astronomer sensed the magnitude of the task in picking out Pluto from the myriads of star images in the background. When Crommelin of England sent his congratulations to the Lowell Observatory, he worded it thus, "Congratulations to you and your staff, (particularly to Mr. Tombaugh)." Dr. Slipher showed it to me.

In the spring of 1930, quite a number of professional astronomers stopped off to visit the Lowell Observatory. It had become the astronomical mecca of 1930. They wanted to see Pluto on the original discovery pair of plates under the Blink-Comparator. Many of them expressed amazement. Relatively few astronomers use a Blink-Comparator in their work—only those who search for proper-motion stars and variable stars. Most had no real concept of what the observational problem involved until they saw the Pluto images on the original pair of discovery plates.

I remember one who, after peering through the Blink-Comparator eyepiece, said to me, "I can barely see the images after you have marked them." I said, "I check planet suspects four to five times that faint all the time."

Another frequent question was, "How did your eyes stand it?" "No problem," I would answer. "I just focus the eyepiece carefully so that my eyes are under no more strain than when looking out of the window. It is the intense mental concentration that wears me down. I take a rest break for a few minutes every twenty minutes, then after an hour, I leave the instrument completely and do something else for half an hour." If one blinks too long at a time, there can be a lapse of attention and one might miss seeing a planet.

I had always assumed that others saw as well as I did. But those many years on the farm looking through my telescope had trained me in how to see. Especially so when one attempted to draw the fine detail on a planet's disk, snatching the best glimpses which occur occasionally during very brief lulls in the atmospheric turbulence.

Astronomers all over the world began to dig back in their old plate files for the star regions photographed in earlier years which might contain the images of Pluto. Even rough calculations would indicate within a few degrees or less just about where they could recover the images. Those plates less far back in time would be easier because there would be less margin of error or uncertainty in position, assuming a circular orbit.

The Pluto image was found on a 27 January 1927 plate taken at Belgium's Royal Observatory at Uccle. This immediately produced the longer arc that would permit a much more reliable orbit. It was computed by A. C. D. Crommelin.

The semimajor axis now bore no resemblance to the earlier Lowell and Lick orbits. The eccentricity was drastically reduced to .287 and the period to 265.3 years—fairly close to the actual values of Pluto. The other elements also were close to the actual values. Altogether, the newly computed elements closely agreed with the ones predicted by Lowell. The status as a trans-Neptunian comet quickly vanished.

Crommelin's orbit greatly facilitated the search and recovery of Pluto images on plates further back in time. Among the first recovered were those on the Lowell plates taken on 19 March and 7 April 1915. Four images were found by S. B. Nicholson and N. U. Mayall on the Mount Wilson plates taken in 1919 for Pickering's Planet O, which at the time were missed.

Mayall remarked that the Pluto images were embedded in the myriad of stars in the Milky Way and that the examination was difficult when covering only a few square centimeters.

To me that seemed like nothing, when I covered that entire Zodiac section of the Milky Way, thoroughly examining over 6000 square centimeters to a fainter magnitude limit, and seeing individually the images of 1.5 million stars!

In the following years, on succeeding strips parallel to the Zodiac, I covered the full width of the Milky Way throughout a length of 220 degrees. This was a plate area of over 60,000 square centimeters, not counting overlap. The regions in Scorpius, Sagittarius, and Scutum were two to four times richer in the number of stars than those in Gemini and Taurus. Considerable portions of these star clouds contained over 1000 star images per square centimeter. Now you may understand why I got tired of the Milky Way. The blinking was brutal.

Images of Pluto were identified on a dozen prediscovery plates; among these were Yerkes Observatory, 29 January 1921; and in 1927, Harvard; and several in Europe, one as early as 1908.

With position observations of a longer arc of the orbit, several parties computed the elements of Pluto's orbit. They agreed very well and were very close to the latest refined values. Among the first of the better orbits were those computed by E. C. Bower and Fred Whipple at Berkeley and by N. U. Mayall and Seth B. Nicholson at Mount Wilson Observatory. These orbits were in remarkable agreement with the ones predicted by Lowell and one by Pickering before he revised it to a poorer one.

Following this, a terrific controversy among notable astronomers raged for decades. From the perturbations on Neptune, Nicholson and others deduced the mass of Pluto to be 0.8 and 0.9 that of the Earth. Other observations suggested a diameter of about 4000 miles (6440 kilometers)—similar to Mars—with a mass of about 0.1 of the Earth. Because of this glaring deficiency of mass, many challenged the validity of Lowell's prediction—among these were E. W. Brown and Dirk Brouwer of Yale. Could this seemingly good fit of Lowell's prediction be a remarkable accident? A. C. D. Crommelin and Henry Norris Russell did not think so.

In spite of his criticisms, Pickering did say in effect that had it not been for Lowell and his observatory, the new planet might not have been discovered for another 100 years. I remember V. M. Slipher showing me a postcard from Pickering, congratulating us on the discovery.

E. W. Brown at Yale University was a leading authority on celestial mechanics. Before the National Academy of Sciences, Brown contended that it was impossible for anyone to have predicted Pluto at all, that its discovery was "purely accidental."

Others challenged Brown, saying in effect, "How do you account for the remarkable agreement of Lowell's orbital elements, and also that Pluto was found within 6 degrees of the place he predicted in the sky?"

As the titans of celestial mechanics were battling it out about the validity of Lowell's prediction, other astronomers, laymen, and the press were anxiously following every paper and article for new information.

For those readers interested in more details, they are referred to the July, November, and December 1930 issues of the *Scientific American* for the monthly feature articles by Henry Norris Russell of Princeton University. Over the years, from time to time, excellent articles were published in *Sky and Telescope*. Among the latter are "Current Problems of Pluto," by Dirk Brouwer of Yale in the March 1950 issue. There was my own seven-page, illustrated article, "Reminiscences of the Discovery of Pluto," featured in the March 1960 issue. Also, the article, "The Mysterious Case of the Planet Pluto," by Dennis Rawlins, of Baltimore, in the March 1968 issue. These later dates suggest how unresolved the Pluto problem was. Many, many other papers appeared in astronomical journals in the United States and Europe, and doubtless in other countries, also.

All of this rumpus came from my finding two tiny, black, starlike images on a pair of plates. But the implications on the structure and origin of the outer Solar System were enormous. It was both exciting and amusing.

Not since the famous Curtis-Shapley debate before the National Academy of Sciences in 1921, on the status of the spiral nebulae, had the astronomical world been so stirred up as the Pluto problem presented in the 1930s. Of course, the Lowell Observatory people were quite upset with E. W. Brown's criticism of Lowell's theoretical prediction of Planet X. Only a few astronomers felt that they had the expertise in the celestial mechanics field to challenge Brown.

In the January 1931 issue of *Popular Astronomy*, Pickering published a long article entitled, "The Mass and Density of Pluto—Are the Claims That It Was Predicted by Lowell Justified?" Bitterly, Pickering claimed that Pluto was his Planet O. But Pickering had predicted too many other planets with frequent drastic revisions for other astronomers to take him seriously. The majority of these were easily within range of the 13-inch (33-centimeter) search telescope. As I was to learn over the next thirteen years in the resumed search, not one of his other planets existed.

In 1940, Dr. Victor Kourganoff, a doctoral candidate at the Paris Observatory made a mathematical reinvestigation of the Pluto controversy, resulting in favor of Lowell's prediction of Planet X. In the late 1940s, Kourganoff visited me in Las Cruces; but by then, I was not convinced that Brown was wrong. To me, the seemingly small mass of Pluto could not be reconciled for a valid prediction. Moreover, my extensive search had revealed no other significant masses.

In the meantime, Crommelin continued to be impressed by the close orbital agreements and position predicted by Lowell and Pickering and unimpressed with Brown's assessment.

In the Royal Astronomical Society's *Monthly Notices*, Crommelin tabulated the predictions of trans-Neptunian planets by Forbes, Gaillot, Lau, Lowell, See, Todd, and Pickering and compared them with the elements he had derived for Pluto's orbit. From this comparison, he felt that Lowell's and Pickering's had merit and the honors should be shared. If Pickering had stuck to his 1919 prediction, he would have been better off, but he spoiled it by making a drastic revision after 1919.

On the other hand, E. C. Bower of the Lick Observatory found Brown's analysis more acceptable. Bower made quite an issue of Pluto's estimated small mass, taken as 0.1 that of the Earth. Bower's conclusion was that Lowell's prediction was a happy accident, concurring with Brown's. Brown considered the discovery of Pluto as an accidental by-product of the search at Lowell Observatory.

A conclusive verdict would not be reached until I had completed the extensive planet search over 70 percent of the entire heavens by 1943.

With the refined computations in Pluto's orbit in the early 1930s, the property of intersecting Neptune's orbit by 1 astronomical unit caused another stir of intense interest. Would Neptune and Pluto collide at some future time? No. The inclination of their orbit planes would never result in an approach closer than several astronomical units. Heretofore, only a few asteroids were known to cross the orbit of Mars. This was before the time of the discovery of small asteroids that cross the orbits of the Earth and Venus. Because of this property of Pluto's orbit, should Pluto be regarded as a super asteroid? There were lively debates on this question. Immediately, the questions arose, "How did Pluto get that way?" "What was the manner of Pluto's origin?"

As I remember, R. A. Lyttleton of England was the first to propose

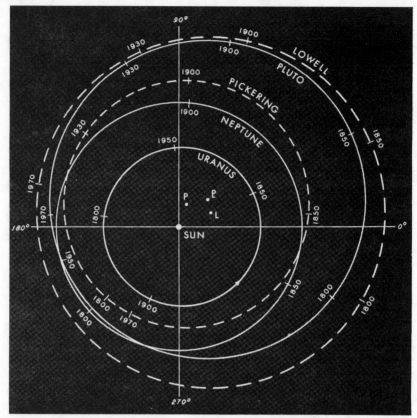

Dashed curves represent the last orbits predicted by Percival Lowell (1914) and W. H. Pickering (1928) for a trans-Neptunian planet, compared with the actual motion of Pluto. The orbits are approximated by off-centered circles, whose centers are labeled. (*By kind permission from* Sky and Telescope, *March 1968, in an article by Dennis Rawlins, "The Mysterious Case of the Planet Pluto."*)

that Pluto may be an escaped satellite from Neptune. Later, several other astronomers proposed or entertained the same idea. As late as 1979, van Flandern developed the idea further, involving the simultaneous origin of Chiron. Chiron appears to be a large asteroid revolving around the sun between the orbits of Uranus and Saturn and cutting across Saturn's orbit slightly. It was discovered by Charles Kowal in 1977 with the large 48-inch (121.9-centimeter) Schmidt telescope on Palomar. Van Flandern proposed that both Chiron and Pluto were once satellites of Neptune and escaped through the violent perturbations of an unknown intruder which has left the scene of the crime. In this supposed encounter, the two known satellites of Neptune had their orbits drastically changed. The smaller satellite, Nereid, has a very elongated orbit around Neptune, while the large satellite, Triton, was changed into a retrograde orbit. These very unusual properties of Neptune's satellites plus Pluto's orbital intersection make it very tempt-

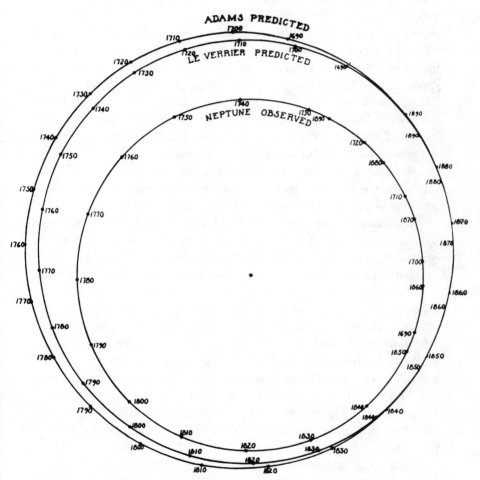

Comparison of Adams's and Leverrier's predicted orbits of a trans-Uranian planet with the actual orbit of Neptune.

ing to consider such a proposal seriously. How long will it take to resolve this riddle?

Still, the striking coincidence of Pluto's position in the sky and the orbital elements to those predicted by Lowell made it difficult for many to accept it as a lucky accident. It was just too incredible! Was there any way that Pluto could have the necessary mass and still retain the faint magnitude and absence of a perceptible disk?

Suppose Pluto was largely composed of the heavy elements—what kind of an answer could be obtained to get the greatest mass possible? From E. C. Slipher's experiments, Pluto could have a disk of .5 arc-second, which at Pluto's distance in 1930, would have a diameter of 8775 miles (14,130 kilometers), as compared to the Earth's polar diameter, 7900 miles (12,721 kilometers). This yields a volume 1.365 times that of the Earth. If it had the same average density as the Earth, the planet would have 1.365 Earth masses. The Earth's mean density is 5.52 times that of water. If Pluto were made entirely out of iron, and allowing for compression of the iron in the center, the density would be about 8.0, and the planet would now have 2.0 Earth masses. Iron is abundant. Indeed, a great number of recovered meteorites are made of iron. So under the most reasonable situation, Pluto seems to be short of the mass required by Lowell by a factor of 3.

Sometime in the mid 1930s, Dr. Dinsmore Alter, my astronomy professor at Kansas University, proposed an idea that might possibly explain Pluto's seeming deficiency of mass. Suppose Pluto's surface was covered with an ocean of liquid air (because of the extremely low temperature that must prevail there). Its curved surface would act as a convex mirror with a radius of curvature equal to the radius of the planet. (Like the convex viewing mirror on the side of one's car.) Incidentally, this would be identical to the virtual image of the Sun seen on the ocean from spacecraft. The temperature on Pluto would be so low that there would be no waves to diffuse the Sun's image. The virtual image of the Sun would be seen only at the apparent center of Pluto's disk. The remainder of the disk would be totally invisible and the actual disk might well subtend an angle of 1 arc-second (as Lowell had expected).

After the Pluto announcement in March 1930, Milton Humason of the Mount Wilson Observatory made a two-hour exposure spectrogram of Pluto with the 100-inch (254-centimeter) telescope, then the largest in the world. With Pluto's light diluted into a spectral band, even of low dispersion, the spectrogram taxed the light-gathering power of this great telescope. In May, Humason paid a visit to Flagstaff and told us that Pluto's spectrum was a solar-type, yellowish, and totally unlike the spectrum of Neptune. This agreed with Lampland's result in comparing the images through blue and yellow filters.

With Alter's proposed explanation, I became interested in what kind of apparent magnitude Pluto would have, if that was the case. So, I proceeded to calculate the magnitude of the Sun's virtual image. Since the spectrum of Pluto was a solar-type, according to Humason, this seemed to be a possibility.

On the first try, I assumed a 5 percent reflection from the liquid surface and a radius of curvature of 5000 miles (8052 kilometers) which would mean a diameter for Pluto of 10,000 miles (16,100 kilometers). The apparent magnitude of the Sun as seen from Earth is − 26.6, the full Moon − 12.2, Jupiter − 2.25. The magnitude scale is such that the succeeding magnitude is the fifth root of 100, or a factor of 2.516. In 5 magnitudes, it is a hundredfold. At Pluto's distance, 40 times the Earth's distance from the Sun, the Sun would be 1600 times dimmer or 8 magnitudes dimmer. To any Plutonian, the apparent magnitude of the Sun would be − 18.6, no longer a visible disk, but an exceedingly bright star point 300 times brighter than the full Moon to us. The effect of the convex mirror surface would dim the virtual image of the Sun enormously more from a distance of 39 astronomical units back to Earth.

From intuitive offhand guessing, I had little idea of what the answer would be. In finishing my calculation, to my utter amazement, the apparent magnitude of the virtual image of the Sun on Pluto's surface came out as + 15, exactly Pluto's apparent magnitude!

A globe 10,000 miles (16,100 kilometers) in diameter would have slightly over twice the volume of the Earth. Allowing for some additional compression of substance material and assuming an iron composition covered by a thin ocean of liquid air, the mass would be about 3 times that of the Earth. But, we need 6.6 to 7.0 Earth masses to produce the perturbations for the residuals of Uranus. It would require a planet larger than 10,000 miles (16,100 kilometers) diameter. This would result in a convex mirror of more gentle curvature and the virtual image of the Sun would be too bright to match Pluto's apparent magnitude.

Amazing! The mystery of Pluto deepens.

Perhaps there are other perturbing masses beyond Neptune's orbit, as yet undiscovered.

13

Decision to Continue the Search

by Clyde W. Tombaugh

As seen from the previous chapter, there was apprehension that Pluto's mass was deficient in fulfilling Lowell's prediction. I noted this uneasiness among my senior colleagues very soon after the discovery, even before the announcement. But they did not openly admit it.

In the course of several weeks, the feeling grew that the real Planet X was not yet found.

I was beginning to get worried. What if the Pluto images were those of a distant satellite of the main body? What if the images of the main body were in the next eyepiece field or a few fields beyond where I had stopped so abruptly? What if some other observatory, in poking around in the immediate area to pick up Pluto on their old plates, had encountered the brighter image of the main body? If this had been the case, this possible blunder could have been tragic for the Lowell Observatory. I should have blinked on a few fields just to play safe. The thought of this possibility makes me shudder yet, after fifty years. I could not rest until I had finished blinking the Delta Geminorum plates to be assured there was no main body within 5 degrees of the Pluto images.

Several astronomers outside the Lowell staff began to urge Slipher that the planet search should continue. They said, in effect, "You have the instrument, the refined technique, and the expertise. What other objects are out there?"

Pluto seemed to be a new kind of object. Were there other similar objects in the vast outer regions of the Solar System? There was only one way to find out—a systematic, thorough search over very large areas of the sky. This was an observational endeavor. Mathematical theory could do no more. There was no favored region in which to look.

Since Pluto was 2 magnitudes or 6 times brighter than the plate limit, the prospect of finding another planet by extending the search area appeared hopeful.

Orbital computation had shown that the plane of Pluto's orbit was inclined 17 degrees and 9 minutes of arc to the plane of the Earth's orbit, the Ecliptic. By accident, Pluto happened to be very near its ascending node at the time of discovery. It could have been midway between the nodes and 17 degrees off the Ecliptic, far outside of the Zodiac belt. It could have been south of the Ecliptic by 17 degrees, or it could have been that much north of it. To have picked up Pluto in such a situation would have required three full plate strips around the sky. Since the Ecliptic is a great circle (equivalent of equator), the area to be searched would have encompassed nearly a third of the entire heavens. There might be another planet with an even higher orbital inclination. This would require more search strips parallel to the others.

One day late in May 1930, after much of the discovery excitement and pressure had subsided, V. M. Slipher came to my office and said, "I would like for you to resume searching. There might be more planets out there like Pluto. Start where you left off in February." The staff now had complete confidence in my work. I had examined two million stars. To cover the wider area, I had at least twenty million stars to go.

Lampland no longer needed the Blink-Comparator for measuring positions, so he removed the delicate micrometer eyepiece, and I put on the old eyepiece which was more suitable for blinking. I was only too glad to place the discovery pair of the Delta Geminorum plates on the Blink-Comparator again. I could do only a limited amount of blinking each day because the night photographic work was on. After about 5 June, the night work was stopped by moonlight. I pressed on with the blinking until the Delta Geminorum pair was finished. It was not an easy pair to do. My star-count samples indicated that about 160,000 star images were recorded on each plate of the pair. This was almost exactly 1000 stars per square degree—and as I was to find later—an almost perfect average per plate. Except for Pluto, nothing more of a planetary nature was encountered on the Delta Geminorum pair of plates except for a dozen or more asteroids.

I don't remember what pair of plates was next placed on the Comparator. Surely not the two pairs in western Gemini, each with one-third of a million star images. It would require several weeks to do them. It was my usual practice to do the thinner plate regions first, especially the more recent ones. It required only three days work to cover each pair with only 50,000 stars, yet covering 160 square degrees. Since another planet could be anywhere, the chances of finding another was proportional to the area blinked. By June, the Gemini region was behind the Sun and inaccessible to anyone.

It is likely that I may have tackled the Rho Leonis plates, because Neptune was in that region at the time. I did work rapidly eastward through these non-Milky Way regions studded with thousands of beautiful spiral and elliptical galaxies and counted the total per section of the plate. Here also, their distribution was uneven and "clumpy."

Then came the plates in Scorpius and Sagittarius, with a million star images each. I had to place a narrow diaphragm 3 millimeters wide and 20 millimeters high in the focal plane of the eyepiece to restrict the field so that I could keep track of the tiny areas scanned. It was a hard day's work to examine a few square inches.

In July and August 1930, I took a longer vacation to visit my relatives in Kansas and Nebraska. Indeed, I received a hero's welcome and gave several invited lectures. Mr. Leslie Wallace of the *Tiller and Toiler* newspaper in Larned and Mr. A. B. Macdonald of the *Kansas City Star* came to the farm and wrote feature stories.

About the middle of August, I boarded the Santa Fe train for Flagstaff to resume the planet search.

In covering adjacent strips around the sky a year later, one must overlap by a calculated amount. A planet could be just barely off the edge of the Zodiac belt and then have moved into it by the following year. A two-year interval would require additional overlap and more

Explanation of the Orbital Planes of the Planets

The dots represent the relative distances of the planets from the Sun. The planets Uranus (U), Neptune (N), and Pluto (P) are beyond the orbit of Saturn at two, three, and four times Saturn's distance from the Sun, respectively. They are marked with arrows, indicating that they are beyond the edge of the chart.

The other planetary bodies are Mercury (M), Venus (V), Mars, asteroids (A), Jupiter (J), and Saturn (S).

The range in distance from the Sun and the orbital inclination for most of the asteroids are represented by the shaded area. The mean distance from the Sun is 2.8 astronomical units conforming to Bode's Law. The largest asteroid, Ceres (C), lies almost exactly at the mean distance. Asteroids are found all the way from Mercury's orbit to beyond Jupiter's.

The orbital inclinations of comets range from 0 to 180 degrees.

The inclination of Pluto's orbit shows why it was necessary to extend the planet search outside of the Zodiac Belt, which has a conventional width of 12 degrees.

Also, the nodes of the orbital planes are widely distributed in longitude, and not in the same direction as shown on the chart.

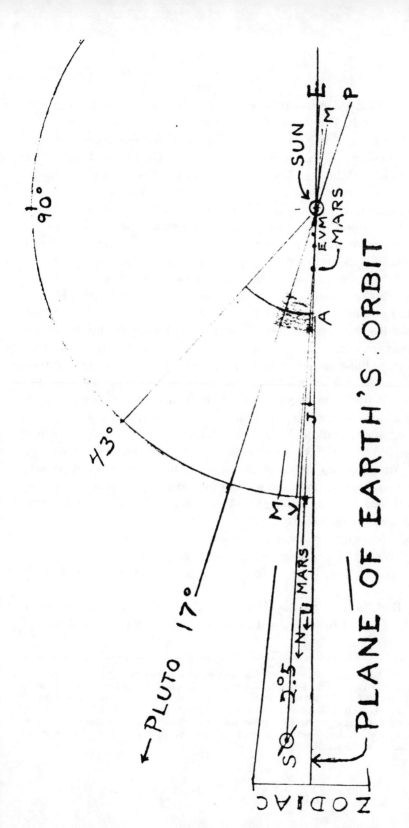

Edge-on view of the orbital planes of the planets inclined to the plane of the Earth's orbit.

work. So I decided to run a "double-header," a strip on each side of the Zodiac belt. The observational photographic work was now doubled, and I had to schedule the work very carefully in order to keep up with the opposition region. It was a fierce pace. Of course, I fell badly behind in the blinking. That did not worry me. Even if the blinking on some pairs was delayed several years, I felt perfectly confident that I could retrieve any faint planet that might happen to be on those plates.

I continued the practice of blinking the thinner regions in order to cover as much area as possible. Any planet that might happen to be in the Milky Way areas would be pretty well concealed and "safe."

With the great interest in possible new planets, the possibility of other searchers getting into the game could not be discounted. Planets could be found in the non-Milky Way regions more easily. That would be the likely places for them to search. So I regarded my record of blinked areas as restricted information until I could get that 35-degree-wide strip finished. Any planet with an orbital plane of higher inclination was a lower probability. After the experience with the greed involved in the Pluto discovery, I could see that even in astronomy, it was a dog-eat-dog situation. After all the sacrifices made by the Lowell family and the staff, I was determined that if there were more planets to be found, they would be found at the Lowell Observatory.

After several months of this pace, the others became concerned. One morning after working most of the night, I got up at about nine o'clock and started working. V. M. Slipher said to me, "Clyde, you should not be up so early."

"I woke up and could not go back to sleep," I said.

"Well, you should lie in bed and rest," he said.

"If I do that, I get nervous. I have a lot of plates to develop," I said.

Early in 1931, the Royal Astronomical Society of England awarded me the coveted Jackson-Gwilt medal and gift of £25 for the discovery of Pluto.

Opposite: Star clouds of the Milky Way in the constellations of Scutum and northern Sagittarius, taken with the 5-inch (12.7-centimeter) Cogshall camera by Clyde Tombaugh on 24 June 1930. Scale: 1.0 centimeter = 1.0 degree. Shown are the lanes and patches of the dark nebulae (vast dust clouds) in front of the star clouds. The 13-inch (33-centimeter) recorded stars 2 magnitudes fainter and about twelve times more numerous. Since this was a two-fold contact to get the picture in negative, the field is reversed. North is at the top, west is on the left.

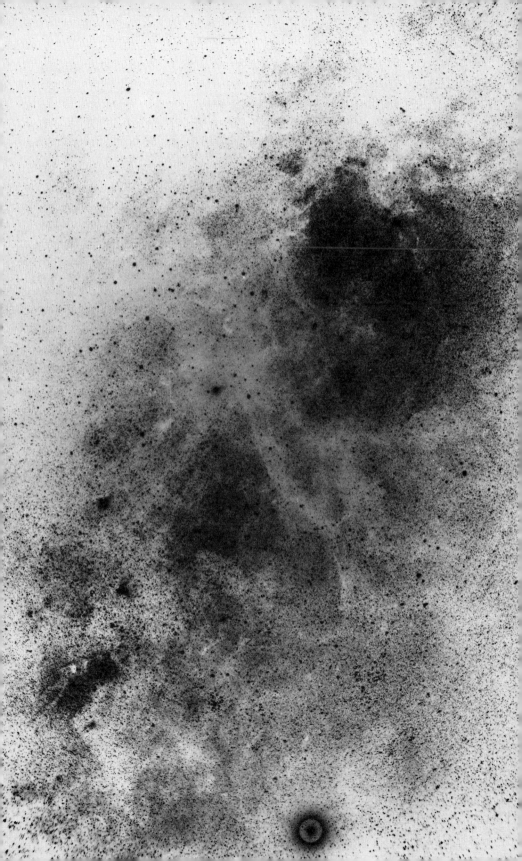

I continued to push the planet search.

In early June 1932, I encountered another major discovery, second only to Pluto. In blinking a pair of plates in the southern strip in a region just south of Libra and just north of Centaurus, I came upon a suspicious object. It looked like a ninth-magnitude star, except that the border of the image looked slightly ragged, instead of the usual soft, amorphous edge for a star of that magnitude. Nor did it have the wider amorphous outline of a distant galaxy. I suspected that it was a very, very distant globular star cluster. There was no partial resolution into stars; it was too far away for that on the 13-inch (33-centimeter) plates. Consulting Shapley's book on star clusters, I found there was nothing listed in that position. Next, I consulted Dreyer's *New General Catalogue of Nebulae and Clusters of Stars*. After correcting the position for precession, there was an object listed at that exact position. The catalog number was NGC 5694, and it described the object as a nebula. But it did not look like a nebula to me.

I called Dr. Lampland's attention to it and said, "I think we have a new globular star cluster. Can you take a long exposure plate with the 42-inch (106.7-centimeter) reflector? It may partially resolve some of the stars."

He did take a plate, and sure enough, no mistake about it. This was the ninety-fourth globular star cluster known to be associated with our galaxy. Herschel had found most of these objects 150 years earlier by visual observation. Dr. Lampland and I published a joint paper of our study of the object in *Astronomische Nachrichten* (Germany) in the August 1932 issue. Many years later, someone else studied the object with a larger instrument, determining its distance to be 103,000 light-years, one of the most distant globular clusters known. We were viewing it just over the impenetrable hub of our galaxy and to the extreme edge on the far side of the galaxy.

In the spring of 1931, I was awarded the Edwin Emory Slosson four-year scholarship at the University of Kansas. The planet search

Opposite: Field of Galaxies in northern Virgo. Portion of a 14-by-17-inch (35.6-by-43.2-centimeter) plate (⅓ of the area) taken with the 13-inch (33-centimeter) telescope by Clyde Tombaugh on 20 April 1931. Same scale as the original: 3.0 centimeters = 1 degree. From a special 5-hour exposure. Faintest stars are of the eighteenth magnitude. The larger and closer galaxies shown are at distances of forty million light-years. For many of the galaxies, the NGC (New General Catalog) numbers were placed next to them. Since this was a two-fold contact to get the picture in negative, the field is reversed.

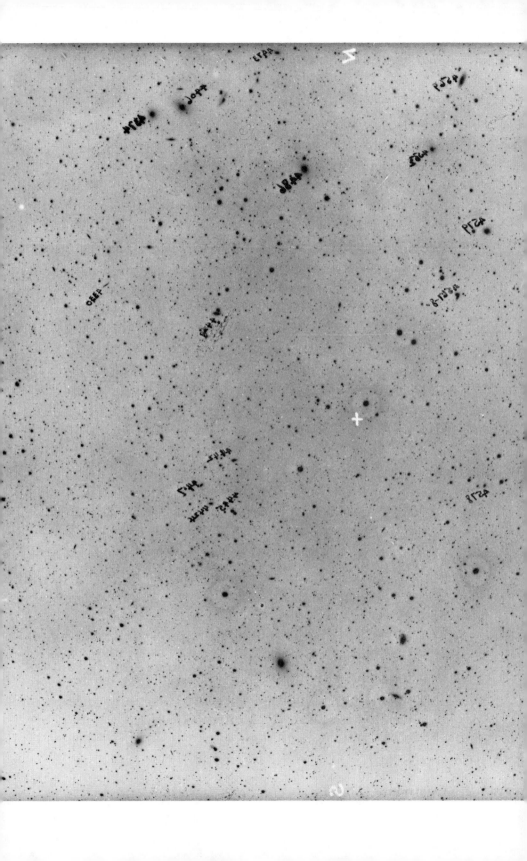

was at such a critical stage that college had to be postponed until the fall of 1932.

In the meantime, the observatory hired a young assistant who was to take plates with the 13-inch (33-centimeter) according to my instructions while I entered the University of Kansas as a freshman.

I had finished photographing the three strips around the sky. Dr. Slipher wanted to extend the search to a still wider limit. The assistant was to photograph the second strip out to the north of the Zodiac. Slipher had me prepare a detailed observing schedule for him.

At last! I was attending a university. I enrolled in the usual courses of mathematics, English, and German. My astronomy professor, Dr. Dinsmore Alter, would not let me enroll in their five-hour course, Introduction to Astronomy. Yet, I had received no prior credits in such a course. He said, "For a planet discoverer to enroll in a course of introductory astronomy is unthinkable." I suppose that it had awkward academic implications.

At the end of the spring semester, I returned to the farm near Burdett for over a week to visit the family. Also, I wanted to finish the short 5-inch (12.7-centimeter) Newtonian reflector I had designed for a visual "Rich-Field Telescope." The focal length was 20 inches (50.8 centimeters). The short focal ratio of 4 made parabolizing difficult, but I soon finished it in the cave, where I did my 9-inch (22.9-centimeter) five years earlier. I took the 5-inch (12.7-centimeter) with its tube to Flagstaff.

On the first day in my office, both Lampland and Slipher, independently, expressed deep concern about the quality of the work done with the 13-inch (33-centimeter) during my absence. Lampland said, "We have been waiting for your return. You have had more experience with these big plates than anyone else." Dr. Slipher wanted me to check all the pairs of plates on the Blink-Comparator for matching quality and any other shortcomings.

It took me over a day to inspect the plates, because so much was wrong. Many pairs were badly matched. For several pairs, the same guide star was not used for the duplicate plate. One pair was particularly bad. I had picked the bright star, Capella, as the guide star for one of the regions. The assistant had used Capella for one plate and a fifth- or sixth-magnitude star some 5 degrees to the south for the other! I don't remember whether he had even made a third plate. On another

NEW GLOBULAR

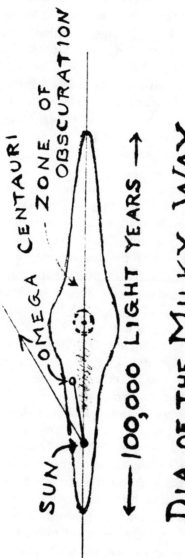

OMEGA CENTAURI

ZONE OF OBSCURATION

SUN

← 100,000 LIGHT YEARS →

DIA. OF THE MILKY WAY

Positional aspect of the globular star cluster discovered in 1932 during the planet search. The diagram shows the edge-view cross section of our galaxy, the Milky Way. The position of the Sun and Solar System is represented by the black dot two-thirds of the radius to the left of the galactic nucleus. The globular star cluster is represented by the small circle on the upper right side. The edge of the Galaxy is not clear-cut, since the stars thin out gradually. Most of the globular star clusters are found within the galactic central 'hub,' or close to it. The "zone of obscuration" is caused by the interstellar dust clouds in the equatorial plane of the Galaxy, producing the dark lane in the summer southern Milky Way (visible to the unaided eye).

pair, something in his observational record made me suspect that the date on one of the plates was wrong. I found an asteroid, obtained its position, and looked through a portion of the asteroid ephemeris. By extrapolating the position from a listed near date, I found the position was wrong on the plate for the date he had recorded.

Now my confidence was shaken regarding the dates and hours he had recorded. In searching for planets, it is critical to know the exact interval between the plates of a pair; otherwise one could be badly misled by the parallactic shift.

Many pairs were badly matched in regard to magnitude limit and unblinkable.

The young assistant came into the Blink-Comparator room several times that day, in a very nervous state of mind. I let him see for himself how badly matched the pairs were. I said, "How could you foul up so badly on that Capella pair?" He shook his head in chagrin.

Upon tallying up the shortcomings, scarcely 25 percent could be used in blinking, assuming that the times and dates were correct.

I took the report to Dr. Slipher and briefly explained the things I found wrong. "That's even worse than I suspected," exclaimed Slipher.

"I can show you examples, if you wish to see them," I said.

"No," replied Slipher, "I'll take your word for it."

Ten minutes after I left Slipher's office, Slipher found the assistant in the hall and I overheard him say, "Here's your paycheck—the last one. I think you know why."

I was disgusted—nine months of time and hard-to-get research money wasted.

It so happened that Frank K. Edmondson from Indiana University had come to Flagstaff to gather observational data on the radial velocities of the globular star clusters for his master's thesis. V. M. Slipher arranged with him to take the planet search plates during the dark-of-the-Moon period each month. He was to test the plates on the comparator for matching but not to blink them.

When I returned the following summer, 1934, I found that Edmondson had obtained an excellent set of plates. In looking through the observational record, I found the notation, "My responsibility begins here. F. K. E."

During the summer rainy season of 1933 I spent nearly all of my time blinking the pairs of plates I had taken in the late spring of 1932. Soon after I returned to Flagstaff in June 1933, Lampland silvered my 5-inch (12.7-centimeter) mirror. The views of the Milky Way star clouds, star clusters, and gaseous nebulosity made a "hit." The faint North America Nebula in Cygnus and the dark "Cat Eyes" in the upper Sagittarius star cloud were quite visible. With this much light power and a field of 2 degrees, there were hundreds of individual stars visible in a single field. Dr. Henry Norris Russell was visiting the Lowell Observatory, and he was ecstatic with the views.

I sent in a description of the views to Albert Ingalls of the *Scientific American*. He replied that to his knowledge this was the first rich-field telescope made in America. He said that S. L. Walkden of London, England, had much the same idea but was using a small, short-focal-ratio refractor.

In 1937, Albert Ingalls edited a new book, *Amateur Telescope Making Advanced*, published by the *Scientific American*. On pages 638–641, Ingalls included my description of the views and a picture of the rich-field telescope. This caught on like wildfire, and in a few years, thousands of such telescopes were made by American amateurs. Many amateurs found this type of telescope ideal for comet hunting.

As I remember, it required another summer's work (1934) to finish blinking all the plates in the first three strips around the sky. Then, I was able to report to Slipher and Lampland that no other planets exist down to magnitude 16.5 whose orbital plane is inclined less than 17 degrees. This absence of planets would also apply to those with higher inclined orbits if they happened to be within 50 degrees of either node.

If we were to extend the search to angular distances greater than 17 degrees south of the Ecliptic, there would have been a problem in the regions south of Scorpius and Sagittarius. It is in these regions that the Ecliptic dips farthest south, where the Sun is positioned in December. Pine treetops blocked accessibility to these regions so low in the south, even when they were on the meridian. I discussed the problem with Dr. V. M. Slipher. We went out to the 13-inch (33-centimeter) telescope dome one afternoon to view the situation. I wanted to go to 50 degrees south of the Celestial Equator. I pointed the telescope to declination 50 degrees south and swung the telescope through an arc of

1 hour on each side of the meridian. We picked out the treetops that were blocking the view. Slipher was in favor of the proposal and said he would try to do something about it.

In a couple days, tree trimmers were on the scene and off came about 20 tons of treetops. No more were cut than was necessary.

Now the planet search could be made over a wide belt within 30 degrees of the Ecliptic all the way around the sky. Of course, as expected, there was a loss of 1 magnitude or a little more because of atmospheric absorption and differential refraction for the regions so near the horizon.

At the end of the spring semester 1934, I married (at the age of 28) Miss Patricia Irene Edson. She was from Kansas City, and we met at Kansas University. We spent our honeymoon in Flagstaff.

As the search progressed to angular distances greater than 20 degrees from the Ecliptic, the number of asteroids on the plates dropped drastically. Dr. Slipher wanted me to do all of the blinking of the pairs of plates. He felt that such a systematic observational search was the only practical approach in discovering any more faint planets.

Putnam and Slipher wanted to get some outside funds to continue the search at a more rapid rate. The blinking work was getting considerably behind. They wanted an additional person to spare me from the night photography work and give me more time for blinking. They applied to the National Research Council for a relatively modest grant, but it was turned down. However, they received two grants from the Penrose Fund of the American Philosophical Society, which gave support for about four years. During the last 2 years, Henry Giclas made the plates with the 13-inch (33-centimeter) telescope.

In April 1938, Dr. Slipher requested a statistical summary of the planet search to date. With this, he prepared a report and read it before the American Philosophical Society on 21 April 1938. At this stage in the planet search, he reported that I had examined 27,000 square degrees, or 65 percent of the whole sky, and had searched through 35 million stars, seeing 70 million star images. Included in the

---

Opposite: Belt and Sword of Orion. This is a portion of a 14-by-17-inch (35.6-by-43.2-centimeter) plate (0.2 of the plate area) taken with the 13-inch (33-centimeter) by Clyde Tombaugh. The scale is the same as the original: 3.0 centimeters = 1.0 degree. The entire constellation of Orion was searched for planets. Not a single star image exhibited planetary motion. North is at top, west on the right.

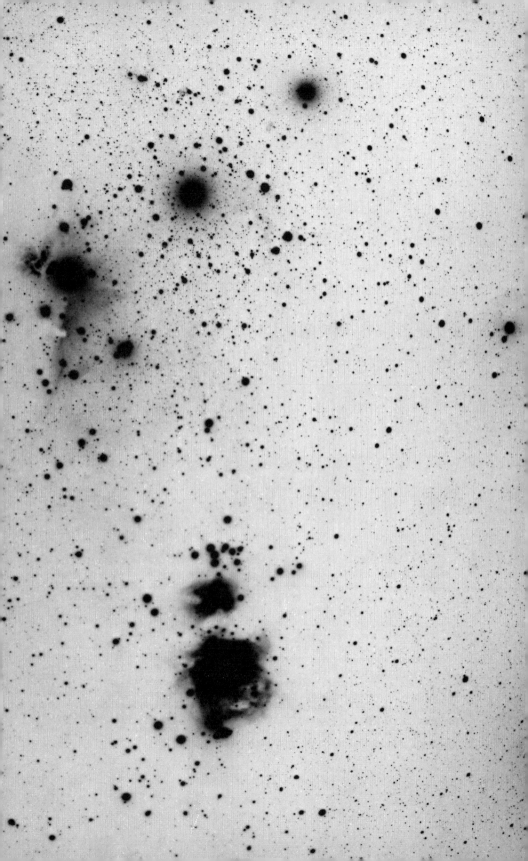

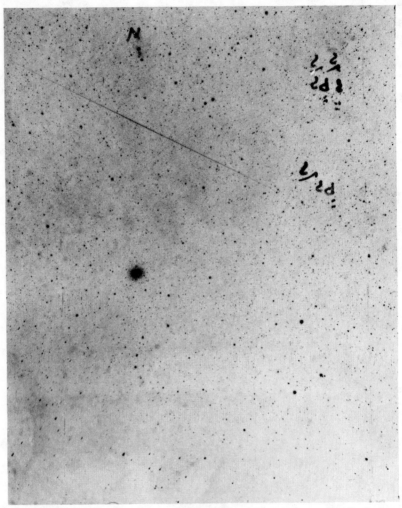

Omega Centauri, largest of some 120 globular star clusters associated with our home galaxy, the Milky Way. Also, this globular is one of the nearest, at a distance of 17,000 light-years. Photographed by Clyde Tombaugh with the 5-inch (12.7-centimeter) Cogshall camera on 7 May 1937, exposure 1 hour, 5 minutes, 47 degrees south of the Celestial Equator, and only 12 degrees N NE of the Southern Cross. Note the meteor trail north of the globular, also the planet suspects (P), which were not confirmed on the third plate. At the bottom of the picture, the stars are blanked out by topped pine trees. Enlarged 2 diameters from the original plate. Scale of picture: 1 degree = 2.0 centimeters. The planet search extended from Omega Centauri northward 110 degrees, covering the area of the Great Dipper (Ursa Major). However, the planet search was made on the large 13-inch (33-centimeter) plates, except for the southernmost strip. For regions at such low altitudes, the star images in the outer portions of the 13-inch (33-centimeter) telescope plates were distorted because the index of atmospheric refraction was appreciably different from the guide star. Omega Centauri is a spectacular object on plates taken with large telescopes located in the southern hemisphere.

report was a sky map showing the position and extent of a great cloud of 1800 galaxies, covering 300 square degrees. This, as I discovered in the autumn of 1936, was located in the constellations of Pegasus, Andromeda, and part of Perseus. All of these galaxies were faint, within 3 magnitudes of the limit of the plates, which would place them at a distance of several hundred million light-years.

The last paragraph was worded thus: "In conclusion, it need hardly be stated in the face of such evidence as has been given that Mr. Tombaugh merits high praise for the keen interest, skill and untiring energy with which he has carried forward this search. Also for making many excellent search plates credit is due Messrs. Frank K. Edmondson and Henry L. Giclas. And finally thanks are due to the American Philosophical Society for its aid of two grants from its Penrose Fund."

During the academic year of 1938–1939, Patricia and I went back to the University of Kansas. In June 1939, she received her Bachelor of Arts degree and I received my Master of Arts degree. Again we returned to Flagstaff.

Under the shadow of impending war in Europe, we continued our work at Flagstaff over the next few years. With the cessation of grant

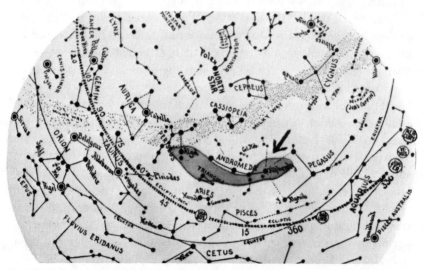

Star chart showing the position and extent of the super-cluster of 1800 galaxies (in the gray shade) discovered in the planet search. The arrow points to the shaded portion, which indicates the position and extent of the Great Perseus-Andromeda Stratum of Nebulae, discovered in the planet search.

money, I returned to the night photographic work. I worked as fast as possible, not knowing when war would stop the search.

During all this search, I kept a sharp lookout for very large asteroids beyond the orbit of Saturn and allowed sufficient overlap to take care of their annual motion toward the Zodiac, even for orbital planes with inclinations of as much as 40 degrees.

Once again, I had some excitement when I encountered a suspect whose shift indicated that the object was about at the distance of Uranus. Upon rephotographing the region, it was not there. I found that circumstances had forced me to photograph a little too far from the opposition point, so it must have been an ordinary asteroid not far from its stationary point.

As the planet search progressed to greater and greater distances from the Zodiac, the prospect of finding another planet became less and less likely.

By now a vast area, extending much outside the plane of the Solar System had been covered. Even the asteroids, some having inclinations greater than 35 degrees, were totally absent.

The war in Europe was worsening. Faced with the uncertainty of how long the planet search could continue, I thought it time to change the strategy. I tackled the non-Milky Way portions of the Zodiac again, prolonging the exposures to two-and-a-half hours. This increased the air-glow, sky-background fogging, but it extended the planet search nearly 1 magnitude, to 18.5. This was the extreme limit possible for the 13-inch (33-centimeter) telescope.

Since the autumn regions of the Zodiac were covered in the fall of 1929 when I had less experience, these were the logical ones to try first. In 1939, I rephotographed the regions in Pisces, Aries, and Taurus. The increased sky fog produced hundreds of very faint planet suspects, all of which were checked with a third plate. It was very laborious, especially the pair in eastern Taurus. The asteroid trails were 2.5 times longer, many of them showing over 1 magnitude variation in brightness within the length of the trail. Also, the longer trails permitted detection of extremely faint asteroids. The number of asteroids I marked ran from 40 to nearly 100 per plate.

By 1940, Pluto had moved eastward into the constellation of Cancer. I wanted to see if there was a group or swarm of lesser planets accompanying Pluto. I made four plates, the fourth as a check plate

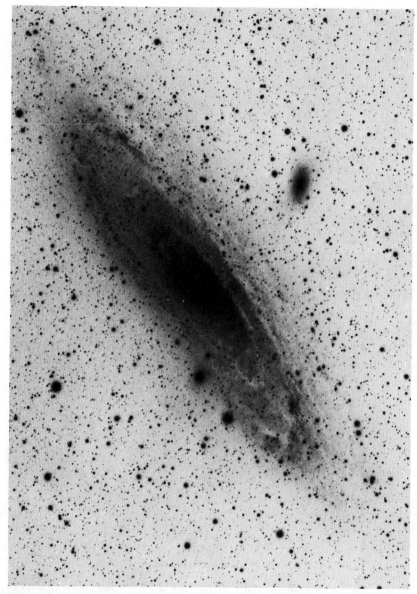

Giant galaxy in Andromeda, M31, at a distance of 2.2 million light-years. Taken with 13-inch (33-centimeter) telescope by Clyde Tombaugh on 17 October 1939. This region was also blinked in search of trans-Neptunian planets.

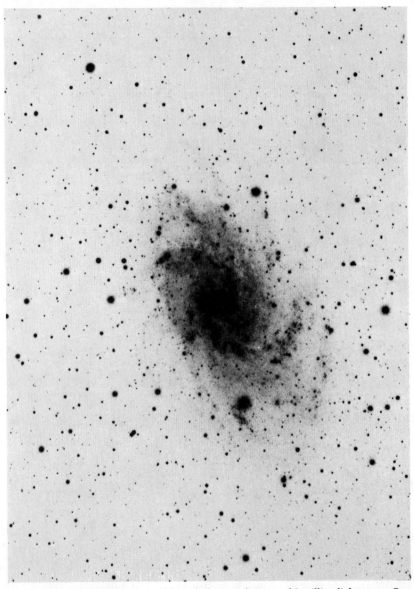

The spiral galaxy, M33, seen nearly broadside, at a distance of 2 million light-years. Our home galaxy (the Milky Way system) would look like this from a similar angle and distance, but twice the size. This spiral galaxy is far beyond the foreground stars of our own galaxy. (Enlarged 3 diameters from a portion of the original plate). The 13-inch (33-centimeter) telescope was not powerful enough to resolve even the super-giant stars of our neighbor spiral galaxy, as the large telescopes do.

for the other check plate. It was on this fourth one that I nearly froze to death in the dome. The sensation of being cold had disappeared. I was so numb that I had great difficulty in closing the dome and getting back to the administration building.

In blinking these plates, I found absolutely nothing in the way of more distant planets. I had thoroughly examined 80 degrees of the Zodiac into the eighteenth magnitude. This was 22 percent of the Zodiac, enough of a sample to cast doubt on the existence of a ring of Pluto-type planets. Therefore, I abandoned my plan to continue the long-exposure search of the Zodiac eastward from Cancer. The long exposures of these Zodiac regions cut in heavily on my other photographic areas. Also, there was the possibility that a more powerful instrument such as a Schmidt camera might be available sometime. If so, a search to the nineteenth or twentieth magnitude of those Zodiac regions would be a significant gain to make it worthwhile.

However, in 1940, I did two plate regions in Libra with prolonged exposures to catch any faint planet before it plunged in front of the rich Scorpius star cloud. Otherwise, one would have to wait three decades before the planet emerged from the Milky Way on the east side. This is also why I did the eastern pair in Taurus.

In the earlier years, I could work at the Blink-Comparator for a total of five to six hours per day. Now my tolerance had dropped to three and four hours per day. The intense mental concentration required in the blinking became more tiring. A few years later, my tolerance in the blinking had dropped to two hours per day. I was "burning out."

Nevertheless, I continued the one-hour exposures to regions still further from the Zodiac as rapidly as possible. The idea was to get the photographic work done without long time breaks in adjoining regions so as to minimize the necessary overlap. The blinking could be done several years later.

By July 1943, the entire area of the sky north of south declination, 50 degrees, had been photographed, from Canopus to Polaris. I had finished blinking all of the accessible southern hemisphere, also all the area between 1 hour to 13 hours of right ascension to north declination 60 degrees (including the Great Dipper).

I had no intention of blinking the north circumpolar area. No planet would have an orbital plane inclination of 60 to 90 degrees. These plates could be used twenty-five or thirty years later as the first epoch set in a proper motion star survey, by someone else.

I had blinked all the pairs of plates to over 35 degrees north of the winter solstice region; much of this was in the Milky Way and very laborious and time-consuming.

I intended to blink a few more pairs in the thinner regions of Corona and Hercules, when the planet search came quickly to a complete stop. I was drafted and assigned to teach navigation in a Navy training school set up on the campus of Arizona State College at Flagstaff until the end of World War II.

My life changed drastically as a result of the war, and my planet searching ended forever.

In fourteen years of work a total of 338 pairs of 14-by-17-inch (35.6-by-43.2-centimeter) plates taken with the 13-inch (33-centimeter) telescope had been blinked, all by me. Also 24 pairs of 8-by-10-inch (20.3-by-25.4-centimeter) plates taken with the 5-inch (12.7-centimeter) Cogshall camera were blinked in the south declination zone between 40 and 50 degrees, except the Scorpius-Sagittarius area which was done with the 13-inch (33-centimeter) because that telescope was much more sensitive to the atmospheric effects at such a low altitude above the southern horizon. The smaller camera was almost as effective, and the plate surface area to be blinked was only one-ninth as much.

Over 90,000 square degrees of plate surface, counting those on both plates of each pair and necessary overlap, were critically scrutinized over every square millimeter, over a total plate area of 75.4 square meters, or 810 square feet. The total sky area is equal to 41,253 degrees; a total of 30,000 square degrees of the sky was blinked. From hundreds of small samples of star counts, the estimated number of stars in the examined areas totaled 44,675,000 (± 1 million), or a total of 90 million star images, counting those on each plate of the pairs. Every one of the 90 million images was seen individually by me. It required a total of 7000 hours of work at the Blink-Comparator.

In this extensive search, the following discoveries were made in order of importance: one trans-Neptunian planet, one globular star cluster, one cloud (or super-cluster) of galaxies, several lesser clusters

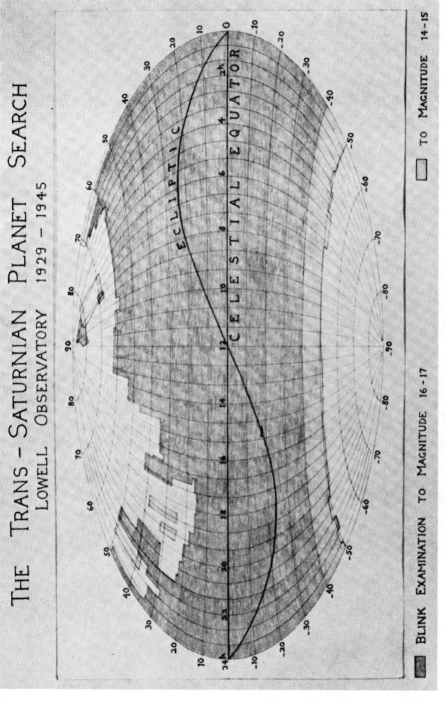

THE TRANS-SATURNIAN PLANET SEARCH

LOWELL OBSERVATORY 1929 – 1945

BLINK EXAMINATION TO MAGNITUDE 16 – 17   TO MAGNITUDE 14 – 15

On this sky map are indicated the regions covered by Lowell Observatory's survey. Nearly the entire sky from declination −50 degrees to +50 degrees has been investigated for faint, distant planets.

of galaxies, five "open" galactic star clusters, one comet, and about 775 asteroids.

On the plates, I marked 3969 asteroid images, 1807 variable stars, and counted 29,548 galaxies. Few astronomers have seen so much of the Universe to such minute detail.

14

Discovery of Pluto's Satellite, Charon

by Clyde W. Tombaugh

In the chapter on "Problems of Pluto," it was seen that the mass of Pluto was a perplexing problem. Various sets of premises indicated a mass ranging from about 1 Earth-mass to .1 that of the Earth. All along, it was seriously questioned that Pluto could be the Planet X predicted by Lowell.

One of the earliest things learned about Pluto was that its light was decidedly yellowish, unlike Neptune. Since the light of the inner planets, the asteroids, and several of the satellites were yellowish, this characteristic seemed to indicate low albedo or reflective power. Low reflective power was a common property of denser bodies like Mercury, Mars, and even the Moon, unlike the large, low-density giant planets: Jupiter, Saturn, Uranus, and Neptune. For decades, Pluto was regarded as comparable to Mars in mass, size, and density.

In early April 1950, I spent several nights at the McDonald Observatory on Mount Locke in western Texas observing Mars with G. P. Kuiper. On one evening of good seeing, we turned the 82-inch (208.3-centimeter) reflector on Pluto. Several magnifying powers were used, up to 3000. Neither of us could see any perceptible difference between Pluto and field stars of the same magnitude. Then Kuiper used his disk-meter and concluded that the disk could not exceed 0.5 arc-second.

Later that year, Kuiper used his disk-meter on the Palomar 200-inch (508-centimeter) Hale telescope and found that Pluto's disk seemed to subtend an angle of 0.23 arc-second. As we know now, Kuiper was not really measuring the diameter of Pluto's disk, but simply the smallest "confusion-disk" of the 200-inch (508-centimeter) mirror. For years, this value was quoted as the size of Pluto, which in 1950 would correspond to a linear diameter for Pluto of 3700 miles (5958-kilometers)—about midway between the planets Mercury and Mars. Kuiper then estimated the mass of Pluto to be about 0.03 that of the Earth. Also, Kuiper entertained the theory that Pluto began its existence as a satellite of Neptune.

Astronomers have to be patient people and await for cosmic opportunities to occur. Eventually, Pluto's disk would occult the light from a distant star by passing in front of it. Since Pluto's disk subtends a very small angle, such an occultation of a star point would be a rare event.

An almost ideal opportunity to measure the angular diameter of Pluto came on the night of 28 April 1965. On that night, Pluto was moving west (retrograde) at about 2 arc-seconds per hour. Ian Halliday of the Dominion Observatory at Ottawa, Canada, had prepared an article of the impending possible occultation in the April issue of *Sky and Telescope*. Since Pluto's apparent path was accurately known, there was a good chance that Pluto would occult a 15.3-magnitude star at right ascension $11^h 23^m 12^s.1$, and declination $+19° 47' 32''$ (1950 coordinates), in the constellation of Leo. The visual magnitude of the star had previously been measured photoelectrically at the Dyer Observatory near Nashville by R. H. Hardie (who had been at the Lowell Observatory in early years). An earlier photoelectric study by Hardie had suggested that Pluto's disk was dark-edged, so that previous diameter estimates had been too small. In 1965, Pluto's visual magnitude was 14.1. The expected time of occultation was about ten o'clock, mountain standard time. Due to some uncertainty as to the exact time, Halliday had cautioned observers to follow Pluto carefully from forty minutes before to forty minutes after the expected event. As

Pluto moved near the star the two objects would merge into one. If Pluto occulted the star there would be a sudden 25 percent drop in the light.

I observed this event visually with my 16-inch (40.6-centimeter) (focal ratio 10) at the Newtonian focus, in Las Cruces, New Mexico. I could see the 15.3-magnitude star without difficulty. As I watched, the star and Pluto merged into one image. The seeing was very good that night, and I could resolve to about .3 arc-second. But I could see no drop in the light. At the end of one-and-a-half hours of such intensive concentration on the observing, I was nearly exhausted.

A much better way to observe such an event is with a sensitive photoelectric photometer. Several observers so equipped followed the event at widely spaced stations. It could have been central for one station to grazing at another. Not one of them recorded an occultation. After the results were studied, it was found that Pluto's disk center missed the star by an amount that proved that Pluto's diameter could not exceed 4225 miles (6800 kilometers) or almost exactly the diameter of Mars. This totally removed the possibility that Pluto might be acting like a convex mirror in which its visibility was due to the virtual image of the Sun at the center of its disk.

Since its discovery in 1930, Pluto has been moving inward toward its perihelion (to occur in 1989). In 1979, it crossed inside of the orbit of Neptune. Pluto is now over 800 million miles (1286 million kilometers) closer to Earth than it was in 1930. This is equal to 9 Earth distances from the Sun. Astronomers had long been awaiting this observational advantage.

In 1974, Michael H. Hart of the Hale Observatories suggested that Pluto might have an atmosphere of neon.

In 1976, another important breakthrough occurred regarding our knowledge of Pluto. By now, Pluto was twice as bright as it was in 1930, permitting special observations through narrow-band filters in the near infrared. In March 1976, Cruikshank, Pilcher, and Morrison, of the University of Hawaii, used the 158-inch (4-meter) Mayall telescope at the Kitt Peak National Observatory in Arizona and found evidence of methane frost on the surface of Pluto. Reflectance measurements between 1.4 and 1.9 microns can distinguish methane frost from water frost and ammonia frost. Methane frost exhibits a deep, narrow-width absorption at 1.7 microns, while the other two kinds of frosts show reflectance maxima at that particular infrared wavelength. Laboratory spectra were taken of methane, water, and ammonia frosts at 77 degrees absolute, − 196 degrees C (320 degrees below 0° F). This was the approximately calculated temperature on Pluto. The available amount of energy through narrow-band filters in the infrared for an object as faint as Pluto is extremely weak, thus requiring the use of a very large telescope. This suggested that the albedo or reflectance was four times higher than had been supposed. A smaller cross section of surface would satisfy the apparent magnitude. This cut the diameter of Pluto to half of the earlier value. Indeed, Pluto may be slightly smaller than our Moon.

Such a drastic reduction in the value of the size of Pluto means one-eighth as much volume and probable mass. Pluto was certainly not Lowell's Planet X.

Pluto is the only planet cold enough for methane to exist as a solid frost. At these trans-Neptunian distances, any other similar body would probably be similarly coated, yielding a higher reflective power and would have permitted detecting other bodies considerably smaller than Pluto. Nothing of the kind showed up in my long planet search. Indeed, a higher reflectance would quite appreciably extend the range of my planet search from the values given in the last chapter, based on a lower reflectance.

If only we had a reliable means of determining Pluto's mass so that we could determine its mean density. Again, the inward travel of Pluto toward the orbit of Neptune provided the observational advantage for the crowning discovery regarding the nature of Pluto.

For several years the U. S. Naval Observatory had been working on improving the orbital data for Pluto. They had been taking plates with the 61-inch (155-centimeter) astrometric reflector at their Flagstaff station. On plates taken on three nights in April and May 1978, Dr. James W. Christy began measuring the exact positions of Pluto. Immediately on 22 June, he noticed a "bump" on the starlike images of Pluto. Since the images of nearby stars in the field were perfectly round, the bump or extension on the side of the Pluto images could not be flaws. The possibility of these being due to any faint background stars nearly in line with the Pluto images was eliminated by examining the 48-inch (122-centimeter) Palomar Schmidt prints. These extensions always appeared on either the north or the south sides of the Pluto images. Consulting their older plates taken in 1965 and 1970, Christy found similar extensions.

There was only one logical explanation: these bumps or extensions were the images of a satellite very close to Pluto! The separation of the centers of the Pluto and satellite images were never more than 0.8 arc-second, and very good observing conditions with a large telescope were necessary to prevent the merging of the two images into one. Robert and Betty Harrington, working with Christy, requested that John Graham examine the Pluto image with the 158-inch (4-meter) telescope at Cerro Tololo Inter-American Observatory in Chile, and the same effect was confirmed. Then similar elongations were reported on McDonald Observatory plates taken in 1977. Also a visual observation of the extension was noted on Mauna Kea in late July 1978.

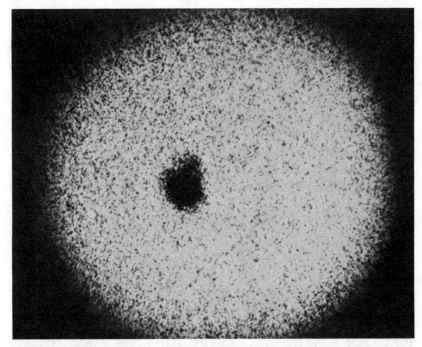

A one-hundred-fold enlargement of the best image of Pluto, taken from a 155-centimeter plate. The satellite is the slight bulge just to the right of the top of the image of the planet. (*Courtesy* Mercury Magazine, *Jan./Feb. 1979.* © *Astronomical Society of the Pacific.*)

Robert Harrington decided to calculate an orbit for Pluto's satellite that would fit the times of the observed extensions.

The separation of 0.8 arc-second at the distance of Pluto corresponds to a linear separation between the centers of the two bodies of only 12,500 miles (20,000 kilometers). Moreover, the period of revolution was equal to the rotation period of Pluto on its axis, namely, 6.39 Earth days. The latter was determined from a periodic variation in brightness of 10 percent soon after its discovery. Over the years, the amplitude of variation has slowly increased to about 20 percent, but the period has remained exactly the same. This variation in brightness is probably due to unequal reflectivity on different sides of the planet presented toward Earth as it rotates on its axis. But the increase in the amplitude of brightness implies that the axis of Pluto is highly inclined from the perpendicular of its orbit plane—similar to Uranus. The behavior of the presentation of the bumps indicate that the plane of Pluto's satellite is seen nearly edge-on, and only at apparent elongation are the bumps visible from the main image. Pluto's equatorial plane probably coincides with the orbit plane of its satellite. This together with the synchronous period would mean that Pluto's satellite would remain over the same place on Pluto, assuming a circular orbit. Such a situation is stable.

The "bump," or extension, on the Pluto image appeared to be about 2 full magnitudes, or 6 times, fainter than the main Pluto image. This means that the diameter of Pluto's satellite is about 40 percent that of Pluto itself. Assuming comparable composition, reflection, and density, this would mean that Pluto's satellite has 0.067 the mass of Pluto. This makes Pluto's satellite the largest one in the Solar System in comparison to its planet. Prior to this, our Moon had that distinction with a mass 0.012 that of the Earth.

Now with the knowledge of the radius of the satellite orbit (adjusted to barycentric because of the appreciable mass ratio of the satellite) and its period of revolution, the combined mass can readily be calculated with the use of Kepler's Third (Harmonic) Law. We get the shockingly low mass of 0.0026, or about 1/400 that of the Earth. Subtracting the probable mass of its satellite, the main mass (Pluto itself) is 0.0023 that of the Earth, according to Harrington. Running through the calculation myself, I got essentially the same answer; my value was

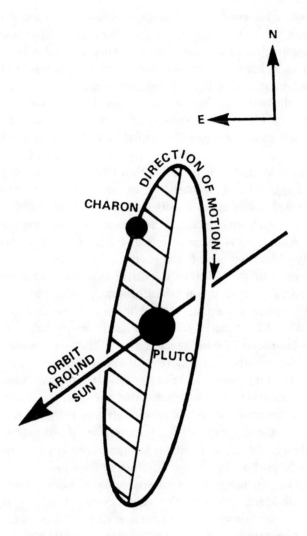

A diagram of the most probable orbit of 1978P1 with respect to Pluto, showing the line of intersection with the plane of the sky, the direction of motion of the system in its orbit around the Sun, and the approximate sizes of the planet and satellite to the same scale. (*Courtesy of* Mercury Magazine, *Jan./Feb. 1979,* © *1979 Astronomical Society of the Pacific.*)

very slightly less. It is difficult to measure precisely the separation of the centers of the two images.

This is a far cry from the earlier value of 0.1 mass of the Earth based upon the assumptions of a lower albedo (reflective power) and a higher density from silica composition. This new value for the Pluto-system mass is hopelessly inadequate to produce appreciable gravitational perturbations on Uranus and Neptune.

With the more reliable new values for the mass and size of Pluto, we divide the mass by the volume and obtain the astonishingly low mean density of about 1.0. This is equal to that of water or compacted ice.

Thus, it seems highly probable that Pluto and its satellite are two huge icebergs, the ices consisting of solid water, ammonia, and methane. Such a low mean density is indicative of very little of the silicates in their makeup. Since methane has a very low temperature sublimation point, − 184 degrees to − 194 degrees C there may be the possibility of another sensational event.

One can calculate the theoretical temperature on Pluto on assumptions of various physical conditions on that planet, but it is near the sublimation point of methane when Pluto is at its perihelion to the Sun (1989). Wouldn't it be interesting if astronomers began to observe methane-gas absorption in the spectrum of Pluto in the next two decades as it warms up a little more while Pluto is inside of the orbit of Neptune? In other words, will the methane frost be transformed into a methane atmosphere? If so, then after the year A.D. 2000, Pluto's methane atmosphere would be precipitated as frost again on its surface, with a winter season lasting for over 200 of our years!

A planet which has a critical temperature range for some given substance to exist at one time or another in the solid, liquid, and gaseous states has the means to generate weather. On Earth that substance is water. On Mars, it is carbon dioxide (dry ice). On Mars, it is rather drastic because carbon dioxide dominates its thin atmosphere. During the winter hemisphere, a considerable portion of its atmosphere precipitates out as carbon dioxide frost on the Martian surface, causing a marked drop in its atmospheric pressure during those seasons. As radioed to Earth by one of the Viking Landers on Mars, the atmospheric pressure is inadequate to permit carbon dioxide to exist in its liquid state. A pressure of 5 Earth-atmospheres is required and so carbon dioxide goes from solid to vapor even on Earth—hence "dry ice." Some of the large satellites of Jupiter appear to have a density almost as low as that of Pluto. Indeed, it is fairly consistent with that of the giant planets. In other words, the outer Solar System is dominated by substances of lighter weight, in contrast to the heavier composition of the inner planets.

Pluto has proved to be the strangest object in the Solar System, and one of the most interesting objects. Because of its faintness and great distance, it required forty-eight years to ascertain its nature.

15

Pluto as a World

by Patrick Moore

Pluto is a twilight world. Admittedly it is now closer to the Sun than Neptune can ever be, but at its aphelion it moves out to 4,580,000,000 miles (7,375,000,000 kilometers), so that it is then at the very frontier of the known planetary system.

Obviously, we can find out nothing about the surface features by direct observation. According to my own estimates, the magnitude is almost exactly 14, so that Pluto can be seen with the aid of a telescope of 10 inches (25 centimeters) aperture or more; the value often quoted in books (14.9) is the mean opposition magnitude, but remember that Pluto will reach perihelion in 1989, so that at present it is almost at its brightest as seen from Earth. Unfortunately, no really measurable disk can be made out even with giant telescopes. When the late Gerard Kuiper made a measurement of the diameter in 1950, using the Palomar reflector, he arrived at a value of 3600 miles (5800 kilometers), which now appears to be far too great. Under these circumstances, there is no need to stress the impossibility of making out any surface detail.

What, then, is Pluto really like? We must now couple Pluto with Charon, because it is indisputable that the pair must be regarded as a double planet rather than as a planet and a satellite (which also seems to kill the theory that Pluto used to be a satellite of Neptune). The temperature is always very low; perhaps − 230 degrees C is a reasonable estimate. Unquestionably the surface is icy, and there is no reason to doubt the accuracy of the 1975–76 observations which indicated a coating of methane frost. This work indicated a higher albedo than had been previously assumed, and hence a smaller diameter of around 1800 miles (3000 kilometers), less than for any of the Galilean satellites of Jupiter. Both conclusions have been borne out by studies of the movements of Charon.

If the combined mass of Pluto and Charon is so low, we have to admit that we are dealing with a new type of body. No longer are we able to class Pluto as a normal planet. There may be some points in common with the icy, low-density satellites of Jupiter (Ganymede and Callisto) and the inner satellites of Saturn; it is less likely that Pluto resembles the nucleus of a comet, because comets have in general very small nuclei, and their masses are much lower even than that of a junior asteroid. Pluto and Charon simply do not appear to fit in anywhere. They are the enigmas of the Solar System.

What we do not yet know, of course, is whether any other bodies of the same kind are moving in those faraway regions. When Pluto next reaches its aphelion (in the year 2114) it will be much fainter than it is now, because it will be more than 1.5 times as remote. If there happened to be "another Pluto" now near aphelion in its orbit, the magnitude would be very faint indeed. At Pluto's aphelion the magnitude will still be 16.5, and as Clyde Tombaugh's searches extended down to 17.0, it would have been detectable; but any planet below the seventeenth magnitude would have been beyond the range of any systematic search yet made. The discovery of Chiron, Kowel's strange trans-Saturnian asteroid, is an extra indication that our knowledge of the outermost part of the Solar System is still relatively meager. Therefore, it is too early to claim that the Pluto-Charon system is unique. All we can say is that it is unique *so far as we know at the moment*, which is not the same thing.

If Pluto's rotation period is 6 days 9 hours, as seems to be the case, then there are over 14,000 Plutonian "days" in each Plutonian "year."

From Pluto, the Sun would appear as nothing more than a brilliant point, though admittedly it would still cast 300 times more light than the full Moon sends to us. Charon, a mere 12,000 miles (20,000 kilometers) away, would loom large—but dim; it would be seen in the guise of a faintly glowing mass, six times the size of the Moon as seen from the Earth, and motionless in the sky. The stars would shine forth brilliantly, but the planets of the Solar System would be barely visible. Even Neptune would be inconspicuous; Uranus even more so, while Saturn and the others would be almost lost. Remember, Pluto is much farther away from Saturn than we are. Comets, too, would cut poor figures. A comet is at its best only when relatively close to the Sun; when moving in the wastes of the Solar System it will have no tail and will be little more than a flimsy ghost.

As for the surface features of Pluto and Charon—well, we can do no more than speculate. There may be layers of frozen material; there may be elevations and depressions; there may even be craters, but we cannot tell. One day, perhaps within the next half-century, a rocket probe will send us information from close range. Until then, we must content ourselves with the knowledge we have. At least we have learned more than seemed possible a few years ago. Cold, dark, lonely iceballs, Pluto and Charon are well named after the god of the underworld and his gloomy ferryman.

16

Beyond Pluto

by Clyde W. Tombaugh

In all of this extensive planet search, I scrutinized the plates very carefully and marked several thousand faint planet suspects, which proved to be spurious when checked with the third plate. Over 90 percent of these were very faint, within the last magnitude of the limit of the plate. As a result, I would be willing to guarantee that there were no more distant planets brighter than magnitude 16.5 at the time the plates were taken.

This survey considerably dims the prospects of more planets beyond Neptune and beyond Pluto. I could have picked up a planet like Neptune at seven times Neptune's distance from the Sun. We have to consider the double inverse-square law. The intensity of sunlight on a planet decreases as the square of the distance from the Sun. The brightness of a planet also decreases as the square of the distance from the Earth.

I could have picked up Pluto at 1.5 times its average distance from the Sun. A planet with the size and reflecting power of the Earth could have been detected at 100 astronomical units, or 2.5 times the distance of Pluto in 1930.

189

According to Bode's Law of the spacing of the planets, if there were more giant planets like Neptune, at least two of them would have been within range of the 13-inch (33-centimeter) telescope for detection. It is obvious that they do not exist. Other planets like Pluto do not appear to exist out to a distance of 60 astronomical units.

The time interval between the plates of each pair taken with the 13-inch (33-centimeter) telescope was never less than two days. This interval was sufficient to permit a detectable shift for an object out to a distance of 500 astronomical units. Jupiter would have been detectable to a distance of 470 astronomical units. For many of the pairs of plates, the interval ranged from three to six days.

The majority of planets predicted were within range of the 13-inch (33-centimeter) and these planets simply do not exist. One time, Kuiper said to me, "The finding of Pluto was an important discovery, but what you did not find out there is even more important." I agreed. The extensive planet search yields important implications regarding the structure of the outer portions of the Solar System and its origin.

As the planet search over a wide belt around the sky was approaching completion, I began to think of reexamining the thinner star regions nearer to the Ecliptic to a fainter magnitude limit. The two-and-a-half-hour exposures over a portion of the Zodiac with the 13-inch (33-centimeter) telescope presented a serious problem. The longer exposures enhanced the sky-background fogging, and the accidental clumping of the grains in the photographic emulsion produced an excessive number of spurious images near the magnitude limit. Since this effect occurs on both plates of a blinked pair, a great number of false planet-suspects resulted, greatly increasing the work of checking them with the third plate.

In about 1941, I discussed this problem with Dr. V. M. Slipher. I proposed a larger instrument that would permit a planet search to the nineteenth magnitude. I had made a preliminary study of the increased number of stars and plate fields. A larger instrument like the 13-inch (33-centimeter), of the refractor type, presented problems of more loss of light from absorption in the thicker lens components and also enhanced secondary chromatic aberration trouble. It would be costly. I proposed a Schmidt type (invented in 1930) with a focal ratio of 4. Since nearly all of the converging power is accomplished by a concave mirror, there is a negligible amount of chromatic aberration, even when using light of all colors. This would be an advantage for yellow-ish planets. The blue-sensitive plates taken with the 13-inch (33-centi-meter) telescope had cheated Pluto by about 0.6 of one magnitude.

Dr. V. M. Slipher was interested but at first favored a focal ratio of 3½. In astronomical photography, the capacity to record fainter stars is dependent on the square of the focal length, also the focal ratio. A faster focal ratio makes the sky-background fog more intense and prevents the longer exposure to record fainter stars.

Since the 13-inch (33-centimeter) had a focal length of 66 inches (167.6 centimeters), a focal length of about 100 inches (254 centi-meters) was necessary to make a significant gain. This, together with the advantage of the Schmidt integrating the colors of the spectrum, would make it possible to detect a planet like Pluto out to a distance of 85 astronomical units. This is a gain of 25 astronomical units farther from the Sun.

A Schmidt telescope with a focal ratio of 5 would permit photo-graphing stars to a still fainter magnitude limit. But the longer focal length increases the scale and decreases the angular field. With this type of instrument, the plate is in front of the concave mirror and occults a significant portion of the parallel light beam from each star. If one increases the size of the plate to gain a wider field, the occulting loss is greater.

After pointing out these conflicting factors, Dr. Slipher was con-vinced that the size and focal ratio of 4 was about the best possible compromise. Accordingly, Dr. Slipher ordered the optics. But really, the 13-inch (33-centimeter) telescope was near the point of diminish-ing returns. One must look for a needle in a bigger haystack.

World War II stopped the whole project and circumstances made it very difficult to be resumed after the war.

Since the discovery of Pluto in 1930, there were at least three re-ports of the discovery of a tenth planet, but none of them materialized. Two of them were claimed to be observed. Apparently, they were mis-takes.

Of tenth planets reported, the one receiving the most publicity was one by Joseph L. Brady of Lawrence Livermore Laboratory, Uni-versity of California, in 1972. This one was really a theoretical case. Brady found that a planet three times more massive than Saturn at a distance of 65.5 astronomical units would reduce the residuals in the time of perihelion passage of Halley's comet at 7 apparitions. Also, he found that the residuals of two similar periodic comets, Olbers and Pons-Brooks, were improved. He assigned the amazing value of 120 degrees for the inclination of its orbital plane to the Ecliptic. This means that its orbit is retrograde, tipped at an angle of 60 degrees. No other planet revolves in a retrograde orbit around the Sun. Indeed, not a single asteroid has a retrograde orbit.

Brady had predicted its position to be north of the constellation of Cassiopeia. This position was only a few degrees beyond my last search strip. I had photographed the area, but it was not blinked.

Upon hearing of this, I immediately phoned the Lowell Observatory at Flagstaff. I said to them, "I don't think that a planet of these characteristics is likely, but it won't hurt to look. These regions had been photographed in the early 1940s." Brady had calculated that his planet would have a brightness of thirteenth or fourteenth magnitude. In his long, proper-motion survey, Dr. Henry Giclas had a great amount of blinking experience, and the planet would not be too difficult to find. Dr. Giclas blinked several pairs of my old plates in that area and found nothing of Brady's planet.

Then Giclas rephotographed the area with the 13-inch (33-centimeter), blinked his plates, and still found nothing.

The disk of Brady's predicted tenth planet would have subtended an angle of about 3 arc-seconds (larger than Neptune's disk).

It is of interest that my co-author, Dr. Patrick Moore, also made a systematic hunt for Brady's planet with his 15-inch (38.1-centimeter) reflector, both visually and photographically, over the whole area and found nothing. There may have been searches made by others, but I am unaware of them.

After Brady's prediction was published, at least one specialist in celestial mechanics studied the implications of such a massive planet at the distance of 65 astronomical units. Because the angular momentum is proportional to the square of the planet's distance from the Sun and proportional to its mass, the angular momentum of Brady's planet would greatly exceed that of Jupiter. Because of the very high orbital inclination of Brady's predicted planet, its angular momentum would pull the orbital planes of all of the other planets into its plane within a period of about one million years.

So here is an important factor to consider in predicting distant, massive planets. The combined angular momentum of Jupiter and Saturn dominates the Solar System and determines the "Invariable Plane," which is tipped from the plane of the Earth's orbit by only 1.6 degrees.

Comets are very unreliable for purposes of predicting planets. When comets come near the Sun at perihelion passage, as does Halley's comet, the ices in the nucleus start sublimating, or evaporating, causing jets of gas to spurt through holes in the crusty surface of the nucleus. This produces thrust like a rocket engine and can change the orbit appreciably, as well as the time of the next perihelion passage. As a result of the search and the properties of comets, it would appear that this case is negative and closed.

Quite often the question is put to me, "Are there any more planets?" The answer can only be a qualified one. It depends on the size of the planet, its reflective power, and its distance. To find a distant planet depends on the limiting magnitude range of the instrument. With the 13-inch (33-centimeter) telescope, it appears that there are no more larger planets. In regard to smaller planets like Pluto, there may be more beyond the distance of 60 astronomical units. Pluto may be one of a new class of objects existing beyond the orbit of Neptune. The very small mass of such objects, the great distances, and the very long periods of revolution would produce no perceptible perturbations on the known planets to offer clues in predicting the existence of such planets.

The only means of discovering more is by an observational search. The large Schmidt-type telescopes could extend the range considerably. It would be of interest to study the problems involved in a fainter search. I made such a study at the close of my planet search.

Extending the search to one magnitude fainter would increase the range of detection from 60 to 75 astronomical units for a planet like Pluto because of the double inverse-square law. The number of stars to be seen and checked in the alternating views with the Blink-Comparator is then doubled for a given size area in the non-Milky Way regions. The number of stars is increased two-and-a-half times in the Milky Way.

Since the time required for blink examination is approximately proportional to the number of stars to be seen, the number of hours required at the Blink-Comparator is increased from 7000 to about 15,000. This is a rather high price to pay for a gain in range of only 25 percent.

Let us consider extending the planet search to three magnitudes fainter, namely to the twentieth magnitude. This will increase the range of detection for a planet with Pluto's character and dimensions from 60 to 120 astronomical units. This increases the volume of space accessible for exploration eight times. The number of stars to be seen and checked in the non-Milky Way regions is increased five times! But within 15 degrees on each side of the Galactic Circle (or the Milky Way equator), the number of stars increases seven to eight times. Instead of having one million stars in the area of one 14-by-17-inch (35.6-by-43.2-centimeter) plate with the 13-inch (33-centimeter) telescope, the Schmidt would record seven million to eight million stars! To cover as much area of sky as I did would require at least 50,000 hours of work at the Blink-Comparator!

Obviously, a planet search to the twentieth magnitude would have to be limited to those areas of sky greater than 20 degrees from the midline of the Milky Way.

Since Pluto-type planets would have orbital inclinations of as much as 17 degrees, perhaps as great as 25 degrees to the Ecliptic, we have a feasible area for exploration 280 degrees long and 50 degrees wide. This is a total of 14,000 square degrees, or one-third of the total area of the sky. Even for these thinner regions of sky, one would have about 120 million stars to look at, or 240 million star images on both plates of the pairs. This will require 19,000 hours at the Blink-Comparator, not counting the extra plate area for necessary overlap.

If one restricts the width of the proposed belt from 50 degrees to that of the Zodiac, which is 12 degrees wide, the work is cut to one-fourth. But now the sky-area sample is too small to have statistical meaning. One could not state that there are no other Pluto-type planets out to a distance of 120 astronomical units.

As I stated earlier, I could have detected a planet like Jupiter to a distance of 470 astronomical units.

It is known that there are many comets with periods of thousands of years. They are discovered only when they come within the inner portion of the Solar System. From calculating their orbital elements, it is known that they recede to distances of thousands of astronomical

Clyde W. Tombaugh, in 1978, at the powerful 24-inch (61-centimeter) planetary telescope at New Mexico State University. This Cassegrainian telescope has the unusually
long equivalent focal length of 150 feet (45.7 meters) to give large-scale photographs of
the planets, which provided a behavioral record for the spacecraft missions to Mercury,
Venus, Mars, Jupiter, and Saturn. The optical system was designed by Tombaugh. (*Photo
by Chuck Williams*)

units before turning around to return to their perihelion near the Sun.
By the time they recede beyond the orbit of Saturn, they are no longer
visible in the most powerful telescopes. Frequently, new comets come
in near the Sun for the first time in the history of civilization. It has
been estimated that there are hundreds of thousands, if not millions, of
comets roaming through the dark, vast outer regions, but they are
unobservable.

From Pluto's mean distance of 40 astronomical units, it is 270,000
astronomical units farther to the next nearest known observable object
—the nearest star Alpha Centauri, at a distance of 4.3 light-years.
There are only five other stars known within a radius of about ½
million astronomical units.

There seems to be no escape from a formidable task if one attempts a deeper planet search. Are there any takers?

Appendix I

The Ninth Planet's Golden Year

by Patrick Moore

On 18 February 1980 a special Pluto Meeting was held at the New Mexico State University, in Las Cruces. This was the fiftieth anniversary of the discovery of the planet. The meeting was arranged both to summarize the current state of knowledge about Pluto, and to pay tribute to Clyde Tombaugh—which is why this account of it is written by me.

Eighteen papers were presented, and some of the points raised were of great interest. Several speakers gave the latest results about the photoelectric and spectroscopic investigations of Pluto. E. Bowell and K. Lumme, of the Lowell Observatory, confirmed that the magnitude difference between Pluto and Charon is about 1.7; Pluto itself is not a low-albedo object, but is reddish, with a surface which appears to be comparatively smooth. E. F. Tedesco (Lunar and Planetary Laboratory, University of Arizona) considered that Pluto probably has an asteroid-type surface covered with methane frost. There was also a valuable paper from R. L. Duncombe (University of Texas at Austin), who gave a résumé of the determinations of Pluto's mass. The estimates have been revised steadily downward, and the latest estimate is 0.002 the mass of the Earth. The diameter cannot be more than 2370 miles (3820 kilometers), and is probably rather less—which means that Pluto cannot possibly be Lowell's "Planet X." It is not massive enough to cause any detectable perturbations in the movements of giant worlds such as Uranus and Neptune.

L. M. Trafton (also from the University of Texas at Austin) dealt with the possibility of an atmosphere around Pluto. The fact that the surface is covered with methane frost indicates a continuous source of gaseous methane which might contribute to the atmosphere. However, Pluto's low mass and low temperature means that a trace atmosphere of methane would escape in a matter of months, and it follows that there may well be a more substantial atmosphere, made up of heavier gases, which would restrict the escape rate of the methane. Nitrogen would not be adequate, since it would itself escape from Pluto in less than a thousand years, but argon is a more promising candidate. Trafton also pointed out that if Pluto had been made of pure methane (admittedly a far-fetched supposition!) it would have evaporated completely in less than 4800 million years, which is approximately the age of the planetary system.

The internal constitution of Pluto was discussed by M. J. Lupo (Massachusetts Institute of Technology). There are insufficient silicates to lead to a liquid core; the overall density of the globe of Pluto cannot exceed 1.3 times that of water. The most likely composition is 21 percent rock, 74 percent water ice, and 5 percent methane. The outer methane-ice layer may be no more than 18.6 miles (30 kilometers) deep, but there is also the possibility that it extends down for half Pluto's radius; we simply do not know. Tectonic features may be expected on the surface, with new material rising continually from the interior. D. P. Cruikshank (University of Hawaii) compared Pluto with other small bodies in the Solar System, such as the satellites of Saturn and Uranus. Most of these bodies seem to be made up chiefly of ice, but beyond Uranus, methane in the frozen or gaseous state seems to dominate frozen water, appearing as a tenuous atmosphere on Triton (Neptune's major satellite) and Pluto, with apparent condensed deposits at least on Pluto.

The main controversial part of the meeting centered upon the satellite of Pluto, Charon. James Christy (United States Naval Observatory) described the discovery and expressed his confidence that Charon really exists; the surface-to-surface distance between Charon and Pluto has been given as 12,400 miles (20,000 kilometers), though this value is admittedly uncertain. Mutual eclipses may begin about 1985, with an uncertainty of three years either way; the eclipse period will last for five years, and there should be 114 eclipses each year (one every three days). If the magnitude-changes during eclipses can be measured, the real existence of Charon will hardly be in doubt. From the surface of Pluto, Charon would appear as a large, dim globe in the sky, and it would remain stationary, since its revolution period is the same as Pluto's axial rotation period—a case unique in the Solar System. Even better confirmation of Charon's existence would be provided by a photograph showing Pluto and Charon clearly separated; this has not been possible as yet, despite efforts by various investigators, notably J. D. Mulholland, using the Mark II Electrographic Camera with the large reflector at the McDonald Observatory in Texas. Attempts have also been made by H. J. Reitsema and F. Vilas (University of Arizona), who consider that while a brightness ratio of 1 to 6 would be detectable, a ratio of 1 to 10 probably would not.

A completely different view was taken by B. G. Marsden (Smithsonian Astrophysical Observatory, Harvard), who was skeptical about the existence of Charon as a satellite; he regretted that it had already been named, particularly as "Charon" is so similar to "Chiron," the strange object discovered by C. Kowal in 1977, and which moves mainly in the region between the orbits of Saturn and Uranus. Chiron is presumably composed chiefly of ice, and the same may apply to the asteroid Hidalgo, which has an eccentric orbit carrying it from the inner part of the Solar System out almost as far as Saturn. Both Chiron and Hidalgo have been given asteroid numbers (2060 and 944 respectively). Marsden considered that Pluto, Chiron, and Hidalgo were essentially similar, though admittedly Pluto has about three times the diameter of Chiron, while Chiron is in turn much larger than Hidalgo. It might therefore be better to class Pluto as a minor planet, and there was even a number awaiting it: 330. The asteroid numbered 330 had been "lost" ever since its reported discovery in 1892, and probably did not exist, as it had been photographed only twice, and its images could not now be found on the old plates. Marsden also speculated as to whether Pluto and similar bodies could be regarded as cometary in nature.

T. C. Van Flandern (United States Naval Observatory) returned to the theory that Pluto might once have been a satellite of Neptune, and was broken free by the perturbing action of a passing body. The hypothetical passing body would have to have been a planet, or at least a member of the Solar System, with a mass from two to five times that of the Earth, so that it could still be within the Solar System in a far part of its orbit. The disturbance could have thrown Triton into its present retrograde orbit around Neptune, and put the small Neptunian satellite, Nereid, into its very eccentric path, which is almost an escape orbit. Van Flandern also commented that the accuracy of Pluto's position as given by Lowell might not be so much of a coincidence as is often thought. Pluto was 6 degrees away from the predicted position, and Lowell had moreover given two possible solutions, so that the area of uncertainty was relatively large in any case.

A. W. Harris (Jet Propulsion Laboratory, Pasadena) disagreed with Van Flandern's conclusions about the past history of Pluto, for several reasons. First, Pluto's mass is too low to be responsible for the retrograde motion of Triton, and the slow rotation of Pluto is better explained by its satellite than by any former association with Neptune; moreover, the Pluto-Charon system could not remain stable in a closed orbit around Neptune, because tidal friction would separate the two in less than a thousand years. It was much more likely that Pluto is a large planetesimal which escaped accumulation or ejection by Neptune because of the 3:2 orbital resonance which prevents close encounters between the two. (Three orbital revolutions of Neptune take about the same time as two of Pluto.) Pluto could therefore be nothing more than the largest of a whole swarm of bodies of this kind. I commented that the discovery of Charon would, if confirmed, weaken the case proposed by Van Flandern; it would be indeed extraordinary to have a satellite of a satellite.

Bradford A. Smith (University of Arizona) referred to the 94-inch (2.4-meter) space telescope due to be launched in late 1983 or early 1984. This instrument would give new information about Pluto's orbit, mass, density, and possible atmosphere, and it could also be expected to show whether Charon really exists as a separate body, since the limiting magnitude of the space telescope would be as faint as 28 or 29. It was not unreasonable to hope that at least some of Pluto's problems would be solved within the next decade.

Following the all-day papers session, a banquet was given by the New Mexico State University in honour of Clyde Tombaugh. He was officially awarded the University's Regents Medal. The citation read as follows:

> The Board of Regents of New Mexico State University takes great pride in presenting the Regents Medal to Clyde William Tombaugh, discoverer of the planet Pluto, ninth major planet in the Solar System. Awarded on the 50th Anniversary of his discovery in recognition of that universal event and in appreciation of his efforts in establishing an internationally recognized planetary research program at New Mexico State University.

Henry Giclas, of the Lowell Observatory, also announced that Minor Planet No. 1604 (1931 FH), one of the numerous asteroids recorded by Tombaugh during his planet-hunt, had been officially named "Tombaugh" in his honor. In his reply, Clyde Tombaugh said that he had been taken completely by surprise by this announcement; he was deeply grateful. He also commented that at least he now had a piece of real estate that nobody could touch!

It was indeed a great occasion, attended by astronomers from all over the world. It was a demonstration of the affection and respect in which Clyde Tombaugh is universally held. So far as Pluto itself is concerned—well, there are many problems yet to be solved, and much research remains to be done in future years. Pluto is truly a world that has come "out of the darkness."

Appendix II

Numerical Data

Pluto

Discoverer: Clyde W. Tombaugh, 1930
Distance from the Sun:

	Kilometers	Astronomical Units
Maximum:	7,375,000,000	49.19
Mean:	5,900,000,000	39.44
Minimum:	4,425,000,000	29.58

Sidereal Period: 247.7 years (90,465 days)
Synodic Period: 366.7 days
Mean Orbital Velocity: 4.7 kilometers per second
Orbital Eccentricity: 0.248
Orbital Inclination: 17.2 degrees
Diameter: ± 1800 miles (2900 kilometers)
Apparent Diameter seen from Earth: below 0.25 arc seconds
Mean Surface Temperature: about − 230 degrees C
Opposition Magnitude at Perihelion: 14
Rotation Period: 6 days 9 hours 17 minutes
Mean Diameter of the Sun, as seen from Pluto: 49 seconds
Number of Satellites: 1

Charon

Discoverer: James W. Christy, 1978
Distance from surface of Pluto: about 20,000 kilometers
Revolution period: 6 days 9 hours 17 minutes (synchronous)
Magnitude: about 16
Diameter: about 600 kilometers (?)
Maximum apparent separation from Pluto, seen from Earth:
    below 1 arc second

Appendix III

Important Dates in the Story of Pluto

1781   Discovery of Uranus by William Herschel

1845   Calculation of the position of a trans-Uranian planet made by J. C. Adams

1846   Neptune discovered by Galle and d'Arrest, on the basis of calculations by U. J. J. Leverrier

1879   Suggestion of the existence of a trans-Neptunian planet by Camille Flammarion

1905–1907   Search for a trans-Neptunian planet by Lowell at Flagstaff

1914   Renewed searches at Flagstaff carried out by C. O. Lampland

1919   Search made by M. Humason at Mount Wilson on the basis of calculations by W. H. Pickering

1929   Search at Flagstaff recommenced by Clyde W. Tombaugh

1930   Discovery of Pluto by Clyde Tombaugh

1949   First direct measurement of the diameter (10,200 kilometers) of Pluto by G. Kuiper

1950   New measurements of the diameter (5800 kilometers) of Pluto, by G. Kuiper and M. Humason with the Palomar reflector

1955   Rotation of Pluto determined by M. Walker and R. Hardie

1965, 1970, 1971   Photographs of Pluto taken with the 1.55-meter reflector at the U. S. Naval Observatory, Flagstaff

1978   Discovery of Charon, by James W. Christy, from studies of the U. S. N. O. photographs
New estimates of the size and density of Pluto

1989   Pluto reaches perihelion

Appendix IV

Model of the Solar System

by Clyde W. Tombaugh

Imagine this model to be set up on a large level field or desert where there are no obstructions to interfere with the picture.

We shall choose as our scale, the Earth's polar diameter of 7,900 miles represented by one inch.

Then the Sun will be a globe of gas 9 feet in diameter, 1.4 times more dense than water and a surface temperature of 11,000 degrees F. The planets revolve around it at various distances in approximately circular orbits, some more elliptical than others.

Mercury will be represented by a large pea revolving around the nine-foot sun-globe at a distance of 385 feet (approximately the length of a football field); Venus by a one-inch steel ball, 700 feet away (approximately twice the length of a football field); the Earth by a one-inch steel ball 970 feet away (over three lengths of a football field); the Moon by a small pea at a distance of 30 inches in orbit around the one-inch Earth ball. (The densities of Venus and the Earth are the greatest of any in the solar system, and are approximately equal to that of iron.)

Mars will be represented by a cherry or a one-half inch marble at

a distance of one-fourth mile (1,470 feet) from the sun-globe; the asteroids by grains of sand to dust particles beyond Mars; Jupiter by a ten-
inch basketball at a distance of one mile (5,030 feet); Saturn by a
slightly smaller basketball (nine-inch) one and two-thirds miles away;
Uranus and Neptune by indoor baseballs (four inches in diameter)
three-and-one-half and five-and-one-half miles away, respectively.
Finally, Pluto by a pea seven miles away. Thus the known area of the
solar system is represented by the space within a circle 14 miles across.
One light-year would be represented by 11,600 miles.

The nearest star, Alpha Centauri, is a double star, represented by
two globes nine and nine-and-one-half feet in diameter, revolving
around their common center of gravity, their distance apart varying
from two to six miles. The distance is so great that these double star
globes must be placed 50,000 miles away to conform to the above scale.

Betelgeuse, one of the super-giant reddish stars, would be represented by a globe of very thin gas three-fourths of a mile in diameter,
and removed to a distance of three and one-half million miles, which
represents 300 light-years.

Model of the Milky Way and the Galaxies

Let us now greatly shrink the scale of our model until a light-year
is represented by one foot. Our Sun would be represented by a speck so
small that it would be invisible in a powerful microscope. Our 14-mile
diameter solar system is now represented by a tiny circle or dot one-
half millimeter in diameter. The nearest star formerly represented by
50,000 miles, is now represented by four and one-third feet. Our great
galaxy of stars, 200 billion strong, the Milky Way, would be represented by a giant discus 20 miles across and three miles thick! The
nearest spiral nebula, the great Nebula in Andromeda, is twice the size
of our own galaxy, and would be placed 400 miles away, while the
most distant galaxy observed (five billion light-years) would be placed
one million miles away, over four times the actual distance of our
Moon.

Glossary

ACHROMATIC OBJECTIVE LENS. Two different lenses very close together, or even in contact, that unite the different color rays over a major portion of the visual spectrum so that they come to a common focus. With a single convex lens, the blue rays come to a focus at a point a little nearer the lens than the red rays do.

ALBEDO. The percentage of sunlight falling upon a planet, a satellite, or an asteroid that is reflected back into space and therefore makes the body visible.

ALTITUDE. The angular distance of a celestial body above the horizon, measured along a vertical circle through the object to the zenith.

AMPLITUDE. The range of variability such as the light of a variable star or the width of a vibration.

ANGULAR DIAMETER. The angle subtended by the diameter of an object.

ANGULAR MOMENTUM. The momentum of an object spinning on its axis (as in the case of a flywheel or a planet) or in a planet's revolution around the Sun. In the latter, angular momentum is the product of the mass of the revolving body times the area swept by the radius vector and is the basis of Kepler's Second Law, the conservation of angular momentum.

APPARENT MAGNITUDE. The apparent brightness of a body such as a planet or star seen from the distance of an observer. Each succeeding magnitude step is dimmer than the preceding magnitude by an amount equal to the fifth root of 100, equal to 2.516-fold. Five magnitudes difference in brightness is equal to 100-fold, 10 magnitudes difference is equal to 10,000-fold, 15 magnitudes is equal to 1 million-fold.

ASTEROID.   A very small planet in revolution around the Sun. They range in size from 600 miles (1000 kilometers) in diameter down to less than a mile or a kilometer. Thousands of asteroids have been discovered and their orbits computed. Most of the asteroids are found between the orbits of Mars and Jupiter. About twenty-five are known to cross the Earth's orbit and are called "Apollo" asteroids.

ASTIGMATISM.   A defect in an eye, lens, or mirror, in which different diameters of the aperture have slightly different focal lengths, making a precise focus impossible. There is also another form of astigmatism related to distortion of images in the outer portions of the field of view.

ASTRONOMICAL UNIT.   A unit of linear measure of distance, equal to the mean radius of the Earth's orbit, or mean distance of the Earth from the Sun. It is the most appropriate unit involving distances within the Solar System.

ATMOSPHERIC REFRACTION.   A bending of light rays from celestial objects by the Earth's atmosphere, making objects appear slightly higher above the horizon than they really are. The effect is 0 at the zenith and increases more and more rapidly toward the horizon where the effect is slightly more than 0.5 of 1 degree (equal to the apparent diameter of the Moon). This causes a severe effect on wide-angle plates in low-altitude regions of the sky in long exposures, because the index of refraction is varying at different rates in different regions of the plate.

BARYCENTER.   The center of mass of two mutually revolving bodies.

BODE'S LAW.   A geometric spacing of the planets from the Sun that breaks down at Neptune and Pluto. If Neptune were eliminated, Pluto's mean distance would fit in well as the succeeding planet beyond Uranus.

CASSEGRAINIAN TELESCOPE.   Most modern, professional telescopes are reflectors of the Cassegrainian type. They consist of two optical members: a large concave mirror (primary mirror) which converges the parallel rays of light from a star to a point at the focus at some distance in front of the mirror at the upper end of the tube; and a smaller convex mirror (secondary mirror) suspended on the axis at the upper end of the tube, and intercepting the converging beam from the primary, returning the beam at a lesser rate of convergence through a small central hole in the primary to a point focus about 2 feet behind the primary mirror. This type of telescope has many great advantages too numerous to relate here. In contrast to the Cassegrainian telescope, the Newtonian telescope is most popular for its smaller sizes and is especially popular with amateur astronomers. The Newtonian is simple, relatively easy to make, and very effective in performance. A concave parabolic mirror converges the rays to a focus at the upper end of the tube. The rays are intercepted by a small, flat mirror at 45 degrees and shunted to the side to the eyepiece. With large telescopes, astronomers don't relish observing at the Newtonian focus from a small, movable platform suspended 30 to 60 feet (10 to 20 meters) above the dome floor.

CELESTIAL EQUATOR.   A great circle traced on the sky where the plane of the Earth's equator extends and meets the celestial sphere or sky.

COMA (TANGENTIAL).   The most severe optical defect in images in the outer portions of the field of a reflecting telescope. The shorter the focal ratio (faster), the more severe is the defect. In usual focal ratios it is different and more severe than the other defect known as astigmatism in the field.

COMET.   A relatively small mass body composed mainly of solid ices of water, ammonia, and methane, in contrast to asteroids and meteorites whose composition is mainly silicates and metals. Most comets have very long elliptical orbits and spend most of their time in deep-freeze at great distances from the Sun. As they approach the relatively brief perihelion portion of their orbit, the warmth of the Sun causes the outer portions of the ice nucleus to sublimate into gas and into a temporary structure of coma and tail. Their orbital planes range in inclination at random from 0 to 180 degrees; those with an orbital inclination of 90 to 180 degrees are retrograde. Most of the comets appear to be members of the Solar System, but a few are known to have hyperbolic orbits, which means that they are interstellar tramps and swing around the Sun only once.

CONSTELLATION.   A group of stars in the sky whose configuration resembled some animal to the imaginative, early, naked-eye astronomers.

DARK NEBULA.   A blank, dark patch in the Milky Way caused by large, extensive volumes of space containing concentrations of opaque interstellar grains of matter (dust clouds), obscuring more distant regions of stars. They are found mostly in the equatorial plane of the Galaxy. Other galaxies seen near "edge-on" exhibit this common feature.

DECLINATION.   The angular distance north or south of the Celestial Equator of a celestial object. It is equivalent to latitude and is one of the coordinates of position in setting the telescope to a particular object by reading the declination circle on an equatorial mounting.

DIFFUSE NEBULA.   Usually an irregular-shaped patch of glowing gas excited by a hot star or group of hot stars within, such as the Orion nebula; or it is the reflection from an intrinsically bright star on a remnant of a primeval dust cloud out of which the star evolved, such as the small patches around each of the brightest stars of the Pleiades. The two diverse types of diffuse nebulae are readily distinguished by their types of spectra. A few hundred of these were visible on the planet-search plates. They are entirely different from the "extra-galactic" nebulae (now known as galaxies). The latter are thousands of times more distant and numerous.

DISK.   The apparent circle of a planetary sphere or globe seen in profile.

ECLIPTIC.   May be defined in two ways. It is the great circle trace on the sky or celestial sphere made by extending the plane of the Earth's orbit to meet it. It is also the exact apparent path of the Sun through the constellations in the course of one year.

EPHEMERIS.   A table listing the positions of celestial bodies from day to day; it also lists angular aspects of apparent size and orientation of planetary disks, the Moon and the Sun, and so on.

EQUATORIAL MOUNT.   A mechanical structure to carry a telescope. The mount has two axes at right angles to each other, in which one axis (the polar axis) is aligned parallel to the Earth's rotational axis, providing convenient tracking of stars across the sky by the turning of the polar axis by a driving clock.

EQUINOXES.   The intersection points of two great circles, the Celestial Equator and the Ecliptic, which are inclined to each other by 23.5 degrees. The vernal equinox is that position occupied by the Sun on 21 March, and marks the beginning of the spring season. It is from a meridian circle passing through the vernal equinox that the hour circles of "right ascension" are reckoned eastward. It is one of the coordinates of position used on an equatorial mount for setting the telescope to a particular star or region.

EYEPIECE.   A small lens, or set of especially spaced small lenses, to magnify the image of a celestial body formed at the focus of a telescope.

FOCAL LENGTH.   The distance between the objective lens and the photographic plate. The longer the focal length, the larger the scale of images.

FOCUS.   The point where the converging cone of light rays comes to a point at a particular distance behind a convex lens or in front of a concave mirror.

GALACTIC EQUATOR.   The midline of the Milky Way, which denotes the reference plane of all stars, diffuse nebulae, dark interstellar dust clouds, galactic (open) star clusters, and globular star clusters in our home galaxy. The richness of star fields is a function of the latitude from this plane. It is inclined to the Ecliptic by 62 degrees.

GRAVITATION.   That mysterious property of matter to attract itself. The gravitational force between two bodies is proportional to the product of the masses and inversely proportional to the square of the distance between them.

GEOMETRY.   That branch of mathematics which treats shapes and spatial relations of objects, involving coordinate positions of points, lines, angles, surfaces, and solids.

HOUR ANGLE.   The angular distance of a celestial object east or west of the observer's meridian, measured in units of the Earth's rotational time.

HYPOTHESIS.   A tentative conclusion to explain certain facts or phenomena and subject to further tests or new knowledge for verification.

JOVIAN PLANET.   Refers to the four giant planets, Jupiter, Saturn, Uranus, and Neptune, characterized by low densities, thick atmospheres, and a composition dominated by the lightest elements.

KEPLER'S LAWS.   Describe the orbital shape, motion, and period of revolution of planets and asteroids with respect to their distance from the Sun.

LIGHT-YEAR.   A unit used to express or measure great distances, equal to the distance that light travels in 1 year at a speed of 186,000 miles (300,000 kilometers) per second. Distances to objects in the Solar System are in light-minutes and light-hours only.

LIMB.   The round edge of the disk of a planet.

LIMITING MAGNITUDE.   Refers to the faintest stars recorded on a photograph taken with a given telescope.

MASS.   The amount of gravitational material in a body; *not* a measure of the size.

MILKY WAY.   That luminous band of light around the sky produced by the sum total of thousands of millions of faint, distant stars. It is also the popular name of our home galaxy.

MERIDIAN (Celestial).   That great circle on the celestial sphere or sky that passes through the observer's zenith from the true north point to the true south point and divides the east and west halves of the sky.

NOVA.   A star that brightens some 10 thousand times within a few days and then slowly fades to invisibility in a few months time, caused by a very severe explosion in the star.

OBLIQUITY OF THE ECLIPTIC.   The angle of inclination between the Ecliptic and the Celestial Equator, caused by the tilt of the Earth's rotating axis from the perpendicular to the plane of the Earth's orbit, equal to 23.5 degrees, and is the cause of the Earth's seasons.

OCCULTATION.   The phenomenon of one celestial body passing in front of another, such as the Moon temporarily hiding a more distant planet or star.

OPPOSITION.   That direction in the sky exactly opposite from the Sun, or 180 degrees from the Sun.

ORBIT.   The true trajectory of a celestial object revolving around a more massive body.

PARALLAX.   The apparent displacement of an object due to a motion or change in position of the observer.

PHOTOGRAPHIC MAGNITUDE.   The brightness of a celestial object on a blue-sensitive plate or film.

PLANETS.   Known to the ancient Greeks as "wandering stars" because of their perceptible movement in the Zodiac, these are inherently dark, lesser bodies which revolve around the Sun as their immediate primary and are visible by reflected sunlight. There are nine major planets and tens of thousands of very small ones known as minor planets or asteroids.

POLAR AXIS.   As related to a telescope equatorial mounting, it is the axis that is exactly parallel to the Earth's rotational axis and provides the most simple and convenient means of tracking stars. The tilt of the axis from horizontal is exactly equal to the observer's geographical latitude.

PRECESSION OF THE EQUINOXES.   The slow slipping of the vernal equinox westward on the Ecliptic through the constellations. It is the result of the gyration of the Earth's rotational poles caused by the gravitational tidal pulls of the Moon and Sun on the Earth's equatorial bulge. It is similar to the gyrations of a spinning top. The vernal equinox slips all the way around on the Ecliptic once in a period of 25,800 years. This effect makes it necessary for astronomers to revise the equatorial coordinate positions of celestial bodies about every decade for setting a telescope on to a given star. As a result, the signs of the Zodiac have slipped westward from the original constellations bearing their names.

PROPER MOTION.   The angular progressive displacement of the nearer stars in reference to the much more distant stars, caused by the orbital motion of the Sun and stars in revolution around the hub of our galaxy. In using my planet-search plates for the first epoch, Dr. Giclas and assistants at the Lowell Observatory have rephotographed the same plate regions some thirty years later, "blinked" these pairs, and catalogued some 15,000 stars which exhibited displacement.

REFLECTING TELESCOPE.   One in which parallel rays of light from a star are converged to a point image by means of a concave mirror. This type has many advantages, particularly in that all the different wavelengths or colors of starlight come exactly to a common focus, unlike the refracting telescope.

REFRACTING TELESCOPE.   Uses a convex equivalent pair or triplet lens to converge the parallel rays of light from a star to a point image at the focus, but there is a serious residual of chromatic aberration in the larger telescopes of this type. Since the light is transmitted through the lenses, it becomes more difficult to cast large disks of adequate homogeneous quality.

REFRACTION.   The bending of light rays through a lens with curved surfaces.

RESOLVING POWER.   The capacity of a telescope to visibly separate the components of a close double star or to reveal fine detail on planetary disks. Resolving power is proportional to the diameter of the primary mirror or the objective lens.

RETROGRADE MOTION.   The apparent motion westward of an exterior planet around "opposition" time, caused by the more rapid orbital motion of the Earth carrying the observer. It may also apply to the true orbital motion of some satellites.

RIGHT ASCENSION. The coordinate in sky longitude reckoned eastward from the vernal equinox in equally spaced 24 divisions of rotational time. The hours of right ascension are subdivided into 60 minutes, and each of those into 60 seconds of rotational time. The span of each "hour of right ascension" is equal to 15 angular degrees.

ROTATION. Pertains to the diurnal spin of a planet on its axis causing day and night.

REVOLUTION. Pertains to the motion in orbit around the Sun in one year.

SATELLITE. Generally pertains to a lesser body which revolves around a planet in an orbit.

SCHMIDT TELESCOPE. The most significant advance in optics for wide-angle sky cameras in the twentieth century. It was invented by Bernhard Schmidt at the Hamburg Observatory in Germany in 1930. It uses a thin plate of glass with very gentle curvature ranging from convex in the central area to concave in the peripheral area. It is placed at the center of curvature of a deeply concave spherical mirror to eliminate spherical aberration. This optical design totally eliminates the defects of both tangential coma and astigmatism (properties of a parabolic mirror), even for very fast focal ratios of 2 or less. The invention of this very efficient sky camera came too late for the extensive planet search at the Lowell Observatory in 1929 and the 1930s.

SEEING. As used by astronomers, pertains to the distortion and swelling of star images from the Earth's atmospheric turbulence. The degree of severity varies from night to night and to some extent from hour to hour. Many very clear nights are rendered useless with large telescopes because it severely limits the amount of useful magnification or large scale on photographs. It is always worse for regions low in the sky.

SOLAR SYSTEM. Pertains to the Sun with its retinue of planets revolving around the Sun, and includes satellites, asteroids, and most comets and meteoroids.

SOLSTICES. The highest and lowest points on the Ecliptic. Also, the solstices are the points on the Ecliptic having the greatest angular distance from the Celestial Equator, equal to 23.5 degrees. The summer solstice is the point occupied by the Sun on 21–22 June; whereas, the Sun occupies the winter solstice on 21–22 December.

STAR CLUSTER. A rich aggregation of stars in a small patch of sky having a common origin of condensation from a diffuse nebula of gas and dust.

SIDEREAL TIME. Denotes the hours, minutes, and seconds of the right ascension region of the sky that is on the observer's meridian at a given instant. Each guide star has its own catalogued right ascension coordinate which is used to set the telescope on the region to be photographed.

THEORY. A set of hypotheses and laws that have been demonstrated to reliably explain a range of phenomena.

TRIGONOMETRY.   That branch of mathematics which relates the ratios of the sides of a triangle subtending various angles. Spherical trigonometry is more complex but is vital in astronomical calculations and reduction of data.

UNIVERSE.   The connotation of this term is often confused among the general public with galaxy or even the Solar System. Universe comprises everything: the Solar System, the Milky Way Galaxy, hundreds of millions of other galaxies, the quasars, and theoretical domains beyond observational range. The Universe is so vast that it staggers the imaginations of astronomers also.

VARIABLE STARS.   Those stars that vary in brightness in regular periods or irregularly. There are several types with different causes or mechanisms. Eclipsing stars are binaries in which the plane of revolution is edgewise in direction to Earth, in which one of the stars gets in front of the other, with precise regular periods. The amplitude in variation of luminosity is always the same for a given eclipsing star, but for different stars the range is from a fraction of a magnitude to 3 magnitudes. Cepheid variables are yellow supergiants, visible at enormous distances, even in the nearer galaxies, with large telescopes. The variation in light is caused by pulsation in size or volume, considerably changing the surface area of radiation. For most of them, the variation is less than 1 magnitude and produces a characteristic rhythmic light curve, with periods ranging from a few days to several weeks. More than 600 are known in our galaxy. The R. R. Lyrae variables are similar to the Cepheids but are not as bright intrinsically, with periods of less than 1 day. They are found in great numbers in the globular star clusters and in the coronal regions around the hub of the Galaxy. Even these variables have intrinsic luminosities about 100 times that of the Sun. The Long-Period Red Supergiants change brightness by 2 to 9 magnitudes in periods of 80 to 600 days, particularly favoring 330 days. Their surface temperatures vary considerably and are much cooler than the Sun. Irregular variables are red stars changing brightness in a most unpredictable fashion, both in magnitude and in time. Since the interval in time between the pairs of planet-search plates was only a few days, the only detectable variables were the Eclipsing variables and the R. R. Lyrae variables. I encountered hundreds of the latter in the globular star clusters and in the star clouds in the direction of the galactic nucleus.

VECTOR.   A quantity that has both dimension and direction. In planet searching at opposition time, the vector of apparent motion is entirely tangential, and the radial vector is essentially zero, permitting a ready determination of a planet suspect's distance by the Earth's daily parallactic orbital motion.

ZENITH.   The highest point in the sky above the horizon, directly overhead.

ZODIAC.   A belt around the sky, 12 degrees wide, centered along the Ecliptic in which the Sun, the Moon, the planets (except Pluto), and most of the asteroids are always found.

Index

About the Authors

CLYDE W. TOMBAUGH discovered Pluto on 18 February 1930 at Lowell Observatory, Flagstaff, Arizona, and holds the distinction of being the only man alive, and the only man in the twentieth century, to have discovered a planet. Tombaugh has written many magazine articles and is Emeritus Professor of Astronomy at New Mexico State University Research Center. He received his Bachelor of Arts and Master of Arts degrees in Astronomy from the University of Kansas and holds an Honorary Doctorate of Science degree from Northern Arizona University.

PATRICK MOORE, O.B.E., is an author-astronomer with over sixty titles to his credit. Recent books include *Guide to the Moon, The New Concise Atlas of the Universe*, and *Guide to Mars.* Moore was director of the Armagh Planetarium in Ireland from 1965 to 1968, is a member of the International Astronomical Union, the British Astronomical Association, and a Fellow of the Royal Astronomical Society. He holds an Honorary Doctorate of Science degree from Lancaster University in England, and in 1979 was awarded the Roberts-Klumpke Medal of the Astronomical Society of the Pacific.